ALKEMA'S SCRAP MAGIC

How to Turn Your Trash Can into a Treasure Chest

By CHESTER JAY ALKEMA

Associate Professor of Art, College of Arts and Sciences
Grand Valley State Colleges, Allendale, Michigan

PHOTOGRAPHS BY THE AUTHOR

STERLING PUBLISHING CO., INC. NEW YORK

Oak Tree Press Co., Ltd. London & Sydney

OTHER BOOKS BY THE SAME AUTHOR

Alkema's Complete Guide to Creative Art for Young People
Aluminum and Copper Tooling
Complete Crayon Book
Crafting with Nature's Materials
Creative Paper Crafts in Color
Greeting Cards You Can Make
Masks
Monster Masks
Puppet-Making
Starting with Papier Mâché
Tissue Paper Creations

ACKNOWLEDGMENTS

The author extends special thanks to the art education students of the College of Arts and Science, Grand Valley State Colleges, for permission to photograph the many projects which appear in this book, and for their assistance in photographing them. Gratitude is also expressed to the following art professors for their valuable suggestions offered in preparing the manuscript and for permission to photograph examples of their students' work: Beverly Berger, Mike Sweeney, Arthur Blom and Tom Butch. Special thanks also to the following for permission to photograph their creations: Steve Anderson, Illus. 302, 305, 306; Marjorie Koster, Illus. 133–136, Margaret Yuill, Illus. 150–155. Illus. 94 by Russel Palmer. Photo of Chester Jay Alkema by Van Dyke Studios, Grand Rapids, Michigan.

419 Park Avenue South, New York, N.Y. 10016
Distributed in Australia and New Zealand by Oak Tree Press Co., Ltd.,
P.O. Box J34, Brickfield Hill, Sydney 2000, N.S.W.
Distributed in the United Kingdom and elsewhere in the British Commonwealth
by Ward Lock Ltd., 116 Baker Street, London W 1
Manufactured in the United States of America

Library of Congress Catalog Card No.: 76-19769
Sterling ISBN 0-8069-5352-7 Trade Oak Tree 7061-2504-9
5353-5 Library

TERRACE LITHOGRAPHERS PRINTED IN USA

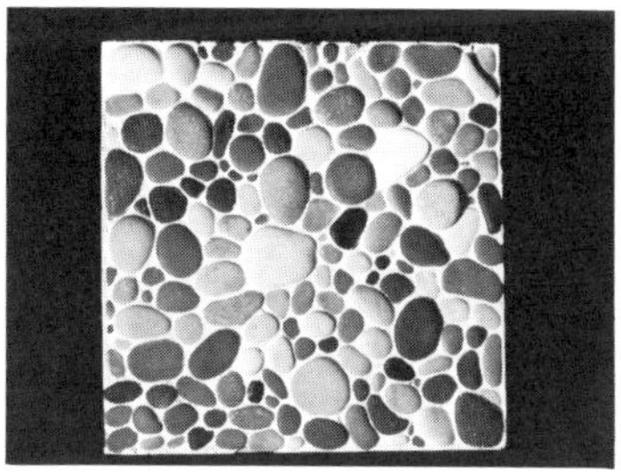

Contents

Before You Begin

Never before have people had such myriads of things to throw away! You will be surprised as you browse through the different sections of this book and think of the number of objects of all kinds that you are discarding every day of your life. At the same time, you might be complaining about the high cost of this article or that article, not realizing that many of the things you buy you can make yourself with the very things you are throwing away!

This book is an *idea* book. It is brimming over with suggestions for making use of all the raw material you have in your own home, basement or garage. The projects are simple, and can be adapted to whatever materials *you* have on hand. The book is arranged more or less in the order of the availability of various materials. For example, paper is probably the most used everyday article in every household, whereas your supply of metal might be more limited. Therefore, we have started with paper and ended with metal. In addition, you need practically no equipment or tools for working with paper, and although you can do a great deal with many kinds of metal using no more than your hands and glue, we have included in that section some more advanced projects involving tools for the ambitious scrap crafter.

So, pick up that piece of paper, plastic container, pine cone, coat hanger, bottle cap, postcard, newspaper, broken toy truck, hub cap, old shirt, wine bottle, or whatever it was you were about to throw away, look it up in the index of this book, and get started!

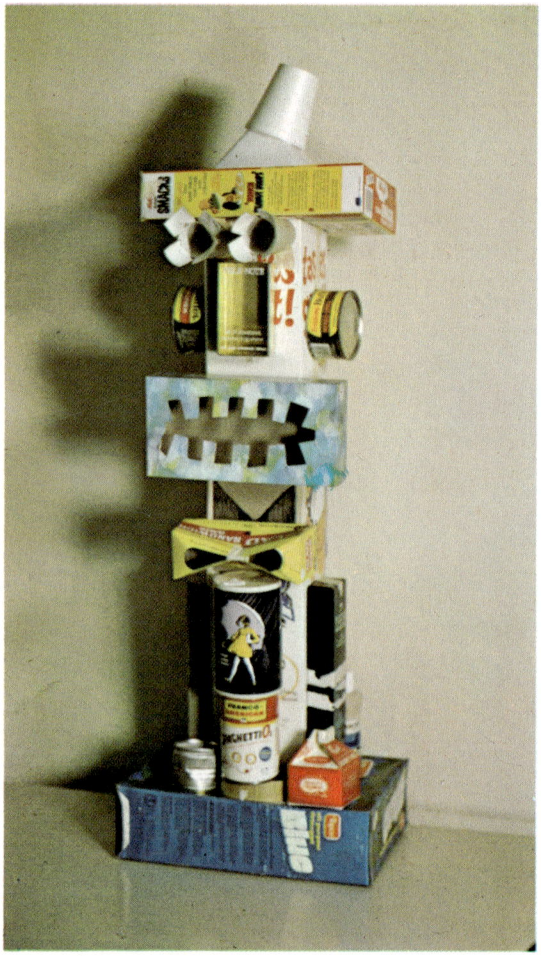

Illus. 1. How many of these objects did you throw away in the last day or two? This robot is just one example of what you can do with discards!

Paper

Of all the "waste" we casually discard each day, none is more abundant than paper products. In one day you might throw away a newspaper, a magazine, a candy wrapper, a milk carton, used wrapping paper, a paper bag, a toothpaste carton, a tissue box, a greeting card, postcard, a shirt cardboard, or a letter and envelope you received (think of all that colorful "junk mail" that goes to waste!). The list is almost endless. Why not recycle these pieces of scrap into something useful, or decorative, or just fun to make?

One of the simplest ways of transforming all sorts of paper is through the collage technique, which is merely the pasting up of various kinds of paper on a background. You can create collages that are purely decorative designs, collages that have specific themes, such as a social comment, collages that are funny or ones that are flights of fancy. Often the paper you

Illus. 2. Here is an example of a decorative collage, which put a sports magazine to good use. The pheasant feathers in the magazine probably inspired the artist to create this fanciful bird. Notice how the solid yellow, brown, and orange elongated shapes complement and echo the colors in the feather design.

Illus. 3. A variety of magazine patterns and colors created this harmonious arrangement of rectangular shapes.

Illus. 4. This exciting collage was created by a basketball enthusiast from a collection of action-packed magazine photographs. By tearing the pictures, and not cutting them, the artist avoided the static effect which sharp edges might have created.

have at hand will suggest the collage you make. Collages can be two-dimensional or three-dimensional, that is, either flat or with raised surfaces.

Shown in the photographs on these pages are examples of collages with specific themes. If you are a sports enthusiast, you might like to choose your pictures from a sports magazine, or if gardening is your thing you can use seed catalogs for ideas, such as in Illus. 8.

Illus. 5. A variety of large and small steps appear to go in different directions, but all lead up to a charming old stone house in the sky. In this magazine collage entitled "Steps," the artist could be expressing a secret dream.

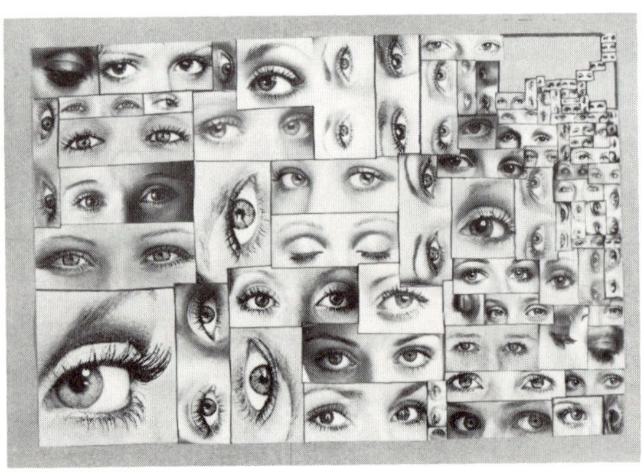

Illus. 6. Pictures of eyes, abundant in women's magazines, provide the basis for this eye-catching collage. When creating a collage of this type, you must be careful to avoid a monotonous effect. Note how the artist here avoided this by using a variety of diminishing sizes.

Illus. 7. This funny fellow really isn't too funny. He symbolizes the pitfalls and temptations that an athlete might encounter in the course of a career. His body is made up of money, his arms are cigars, and he has whiskey-bottle legs. All are cut from magazines of various kinds.

Illus. 8. Don't throw away those seed catalogs and seed envelopes. Look at what one artist created from old catalogs. The flower pictures were cut out of the catalogs in petal shapes and put together to form blossoms of different kinds. The leaves were painted in with tempera.

Illus. 9. This charming collage harks back to a 19th-century Polish folk art. Originally silhouettes executed in one color, these designs evolved into several layers of multi-colored papers. These typically intricate designs can be simplified and constructed of many kinds of found paper. The collage shown here was created from magazine papers in tans, browns and black.

Illus. 10. "Dream Images" is composed of a wide assortment of ghostly faces and demonlike creatures taken from science-fiction magazines. The central figure is surrounded by a bursting blue cloud. Beneath this is a rowboat holding three men desperately fighting a perilous sea.

Illus. 11. "Outer Space" provides the theme for this dramatic collage. A large hand holding space-age gadgetry is surrounded by the infinite regions of the Universe. Scientific magazines and other publications would be a good source of material for a collage such as this.

Illus. 12. The "elongated" collage technique gives you a chance to really exercise your imagination! Here, a picture of a small car was cut up and put together in a unique and effective way.

Illus. 12 and 14 show an exciting method you can use to make your collages. Simply take one picture, and cut it up into a number of strips and other shapes. Then glue them down on a background paper leaving narrow or wide spaces between each piece, but assembling them in such a way that they retain their recognizable original form. The two examples shown were done in two different ways.

The picture in Illus. 14 was merely cut up into many strips and pieces and very narrow spaces were left between them in the reassembling in order to emphasize the elongation of the dancing and jumping figures. The little car in Illus. 12 was done in a more exaggerated fashion, and the various parts of the body of the car were replaced with cutouts which represent scenes you might pass on a trip. The original parts of the car that were cut out were used as patterns for the substitute cutouts.

Another interesting effect can be achieved with the collage technique by weaving paper strips into slits made in a picture cut from a magazine, as shown in Illus. 13. If you want, you can use another complete picture cut into strips and woven into the parallel slits, or you can use a variety of unrelated strips for a more abstract collage.

Illus. 13. Here is another technique you might want to use with your scrap paper—a woven collage. The three small faces were pasted down on a background and the large face was cut into strips. Slits were made in the three faces both horizontally and vertically to receive the weaving of the strips.

Illus. 14. The motion of these dancing and jumping figures is enhanced by the elongation of the forms. Notice how the movement of the racing horse is exaggerated and thus more exciting.

  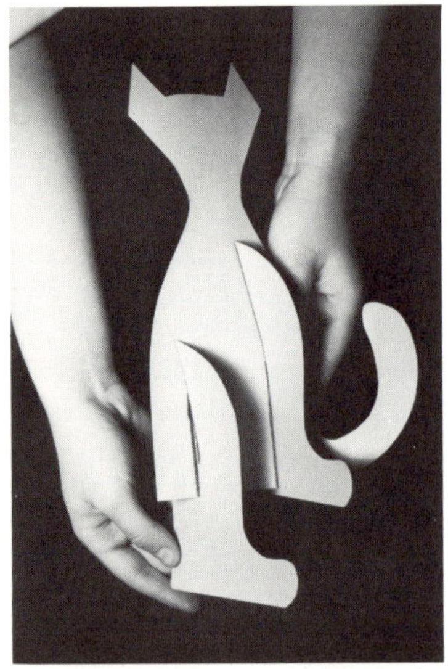

You can use either heavy paper or cardboard for the slit-and-slide technique. (If you do not have paper heavy enough to stand up, paste several sheets together.) Illus. 15 (left) shows a finished cat all ready to decorate a child's room, and Illus. 16 (middle) and Illus. 17 (right) show how it came to be.

The slit-and-slide technique offers you the opportunity to create three-dimensional figures which are ideal for children's rooms. In Illus. 15 a cat shape cut from three pieces of cardboard has a slit cut in each leg and two slits cut in the body. The slits must all be the same length. Then the cat is simply assembled by sliding the parts together through the slits.

Illus. 18. This charming slit-and-slide elephant was created from corrugated cardboard painted with blue and white acrylics, and is all ready to take his place next to the cat in the nursery or playroom.

Illus. 19. Corrugated cardboard is the basis for still another kind of "sculpture." This time—a bas relief wall hanging, all ready for a coat of paint.

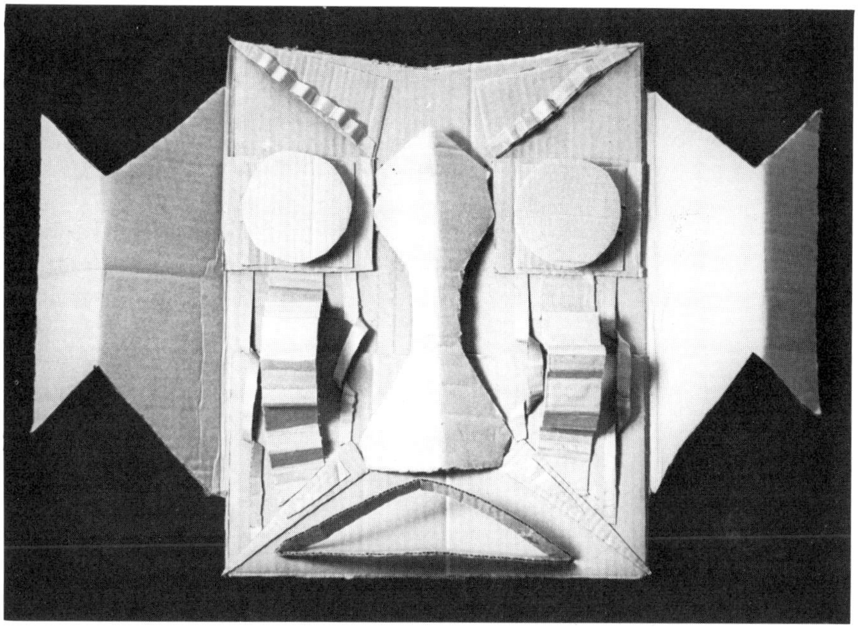

A bas-relief sculpture can be created easily from scrap pieces of corrugated cardboard as the wall hanging mask in Illus. 19 shows. Here, the nose is bent and the corners are inserted into slits cut into the background piece. In addition, pieces can be glued one on top of another to extend shapes from the background, as the eyes are done here. Be sure to paint your bas relief or decorate it with any odds and ends of fabric, buttons, and so on, that you might have on hand.

Illus. 20. An unusual corrugated cardboard creation—an A-frame doll house. Masking tape holds the pieces in place, and pieces of rolled tagboard provide the foundation stilts. Note the small garage at the side of the house. You can paint a doll house such as this with either acrylics or tempera.

Illus. 21. Don't know what to do with all those greeting cards you have left over or have received and just never got around to throwing away? Make a paper mosaic! This happy, dancing figure, made up of green tesserae, is set in a red-tesserae background.

Although you may be most familiar with mosaics made of other materials, all of which require a certain amount of equipment, you will find paper mosaics are fun and easy to do. You can use all kinds of paper for your mosaics. Search for variety, such as contrasts of thin, heavy; light, dark; glossy, dull; rough, smooth; and transparent and opaque papers. Create wall hangings, decorate old serving trays, TV dinner trays, canister sets. Turn coffee cans into decorative wastebaskets, curler boxes, mending kits, or yarn holders. Glue plain paper over the surface before decorating with your small pieces of paper tesserae.

The butterfly shapes in the colorful mobile in Illus. 22 were created by laminating wax paper and tissue paper together. The tissue was cut into the desired pattern and laid on top of wax-paper butterfly shapes. A second, identical wax-paper butterfly shape was placed on top of the tissue and the three layers of paper were ironed together with a warm iron. The heat of the iron melts the wax paper together and thus laminates the tissue between. The mobile parts were hung from twigs with thread.

Illus. 22. If you have any odd pieces of tissue paper and wax paper, you might want to make a brilliant laminated window hanging or mobile such as this one.

Illus. 23 and 24. These elegant, easy-to-construct sculptures are made from a material called foam core, often discarded from architects' offices. This material has a layer of foam sandwiched between two pieces of manila paper. Light and easily handled, it is ideal for sculptures. Pins and white glue serve to hold the pieces together.

Illus. 25. Colorful carton sculptures are ideal for decorating children's rooms or playrooms, and an ideal way to "dispose" of those extra cartons that pile up.

To make the carton sculpture shown in Illus. 25, remove shapes from the carton as shown in Illus. 26. Then, cut shapes from still another carton to fit into the cut-aways in the first carton. Anchor them securely with masking tape. Then attach smaller cartons with tape and glue to the large unit. Decorate the sculpture with tempera or acrylic paint. The large carton shown here measures $1\frac{1}{2} \times 2\frac{1}{2}$ feet (.45 × .76 m.).

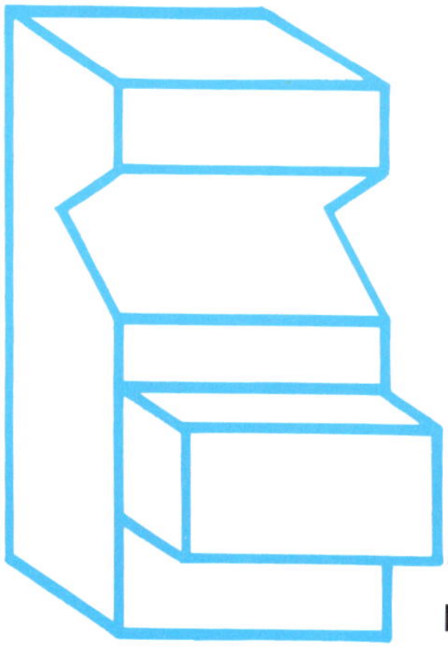

Illus. 26.

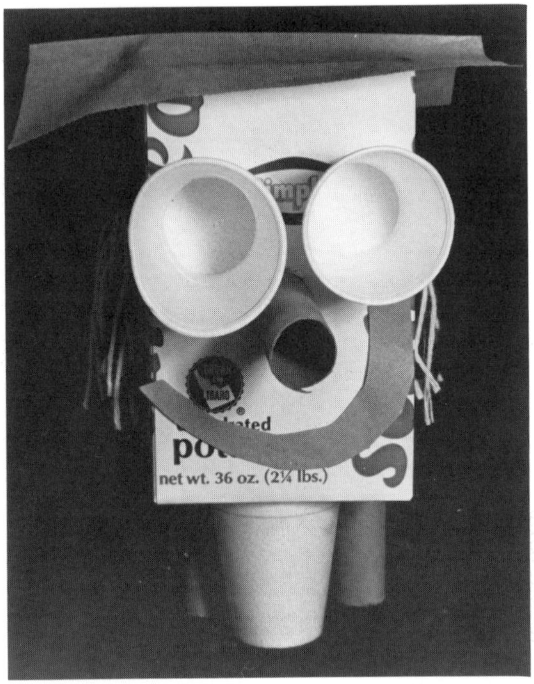

Illus. 27 (left). A comical "person" is just as much fun to make as he is to look at, because you can really stretch your imagination and materials to create one. Here, an instant potato carton forms the foundation for paper-cup eyes and legs and a toilet-paper-roll nose. Paper towelling serves as a hat and moon-shaped mouth.

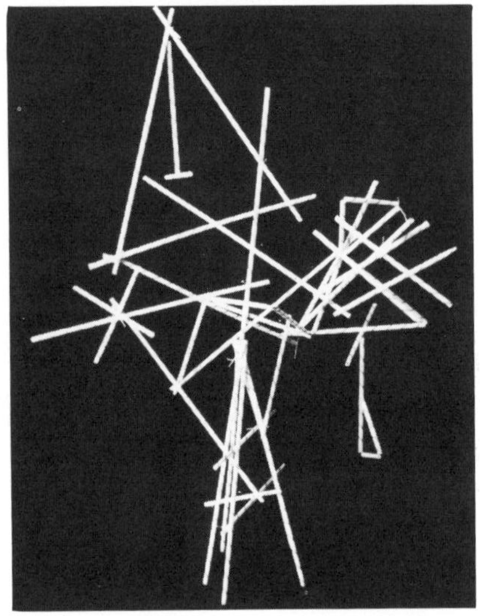

Illus. 28. Today, most paper drinking straws do not disintegrate after use. Simply dry them out, set them aside until you have a large supply, and then construct a stabile. Here, four straws, glued together at the top, make a pyramid upon which more straws are built up. Use straight pins to hold everything together until the glue has dried. You can also stitch the straws together with needle and thread if you prefer. If you use pins, avoid puncturing the straws all the way through so that you will not prick your fingers while handling the stabile. Use colored straws and you will not have to paint them.

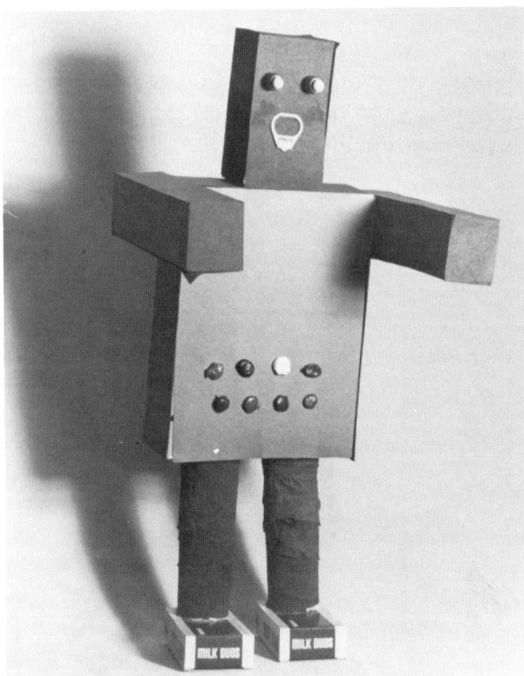

Illus. 29. This little robot is composed of a variety of small cartons, covered with pieces of red and black construction paper and crepe paper. Lumps of melted wax candles (you could also use crayons) make up his control center. Beads, a flip-top cap and tin foil were on hand for decoration, but almost anything similar could be used.

Illus. 30. Folded magazine sculpture allows you to make stunning figures such as "The King." You can turn an old magazine into almost any kind of person, animal, fish, holiday decoration, or whatever else strikes your imagination!

"The King" in Illus. 30 is an example of folded magazine sculpture. Take a magazine and fold each page on a diagonal in towards the center of the magazine. Notice that the diagonal crease ends approximately 2 inches (5 cm.) from the bottom of the magazine. Folding on the diagonal will provide the full skirt shown in the photo. Join the front and back covers with glue to create a circular structure. Rectangular pieces of construction paper were rolled into cylindrical shapes for arms and glued to the body. A cone-shaped construction-paper collar has fluted edges. A plastic-foam ball is glued to the top of the magazine body to form a head.

Before the head and arms were attached, the magazine was spray-painted. Sequins were glued on the construction-paper crown, sleeves, and collar. His facial features were cut from paper

Illus. 31. A "Reader's Digest" and a "TV Guide" were put to good use here to make two charming little people. Notice all the odds and ends that were used to decorate them—buttons, sequins, pieces of yarn, ribbon and fabric.

and his moustache is a dyed pipe cleaner. Yarn is glued on for his hair and beard. Pipe cleaners are bent into hand shapes and attached to the body by running them up through the sleeves and attaching them between the pages near the top of the magazine.

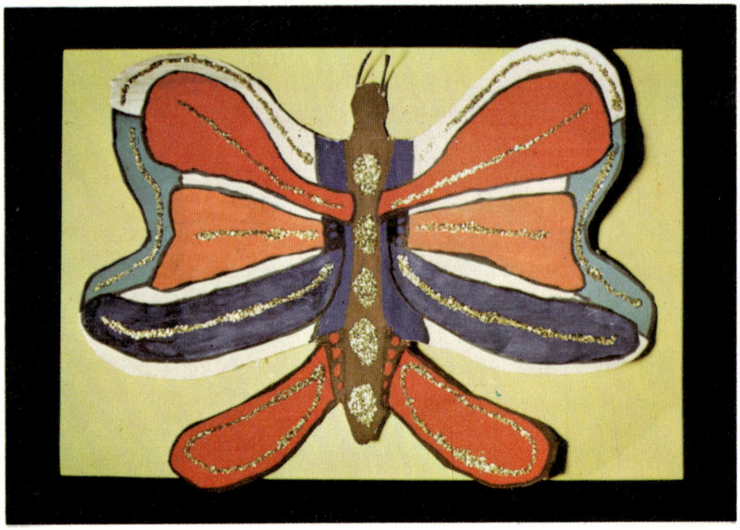

Illus. 32. It's hard to believe that ordinary newspaper lies under this beautiful four-layer papier-mâché butterfly.

The papier-mâché method provides one of the most exciting techniques for your scrap-paper crafting. The butterfly in Illus. 32 was made with the four-layer method, which is one of the simplest ways of making papier mâché. Four sheets of newspaper were pasted together with white library paste. Then the butterfly shape was immediately cut from the paper and the wings slightly curled forward. Crushed newspaper, placed beneath the wings during the drying process, helped retain their bent shape. When dry, the shape is rigid and retains its three-dimensional appearance. Tempera and Christmas glitter were used for decorating the butterfly.

Illus. 33. After your next birthday party, hold onto those balloons! You can re-cycle them for another party by using them as moulds for exotic papier-mâché decorations such as this scary fellow, who is all ready for Halloween.

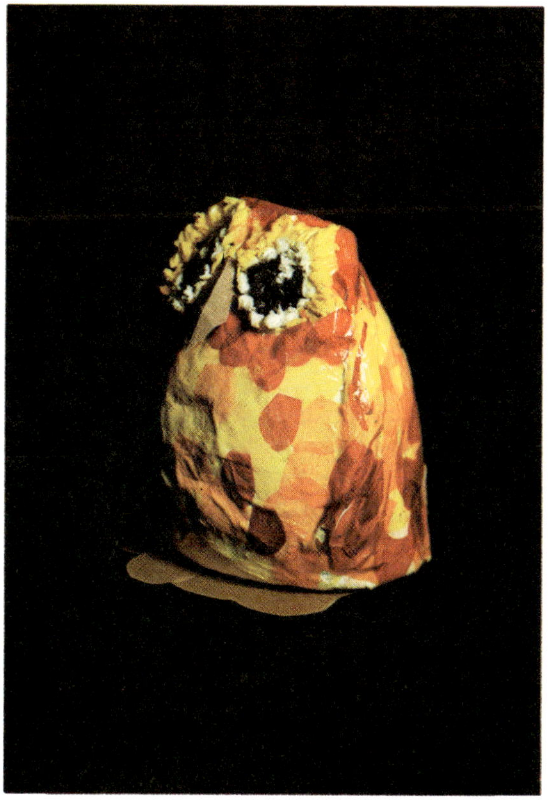

The owl and the mask on this page were made with papier mâché moulded over balloons. After covering the balloon base for the mask in Illus. 33 with four layers of papier-mâché strips, cardboard tubes were added for ears and a cardboard cylinder was added for the base. More strips of papier mâché were used to attach these parts to the head. Paint decorates the finished mask.

Illus. 34. A wise old papier-mâché owl made over a balloon. Tissue-paper feathers are glued to his white-tempera-painted body with liquid starch.

Illus. 35. A wire frame, covered with papier-mâché strips, forms the skeleton of this hippie guitar player.

You can create interesting sculptures by covering wire with papier mâché. Tear old newspapers into strips from top to bottom (that is, going with the grain of the paper). You will need narrow strips for small objects and wider ones for larger objects. Submerge the strips in a flour or wheat paste and wrap them round and round the wire frame until your figure is as large as you want.

The guitar player in Illus. 35 has a wire skeleton covered with papier-mâché strips and painted with tempera.

The blossoms in Illus. 36 are made of papier mâché over a wire frame. The wire was sculptured by hand to form the edge of each petal. The surface is decorated with tempera, tissue paper, Christmas glitter and construction paper.

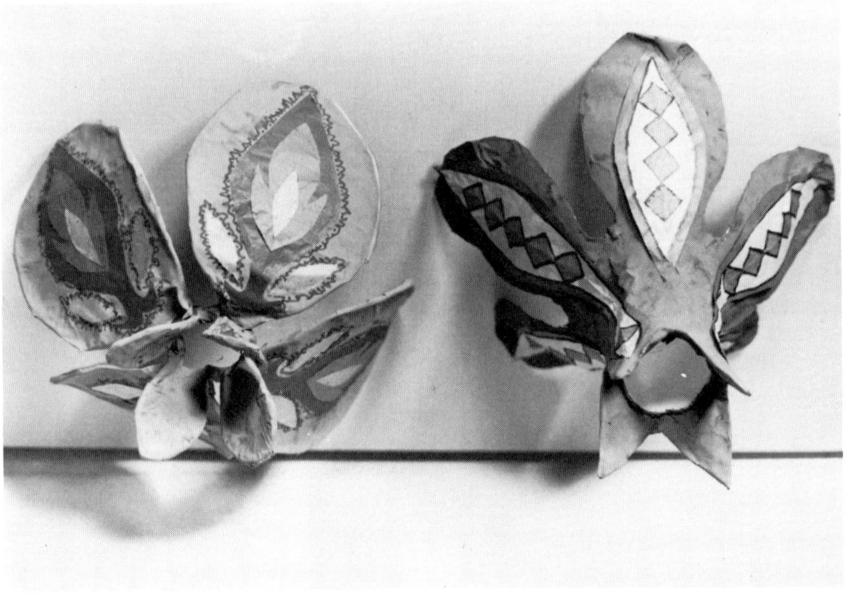

Illus. 36. These "blossoms" are made of papier mâché over a wire frame.

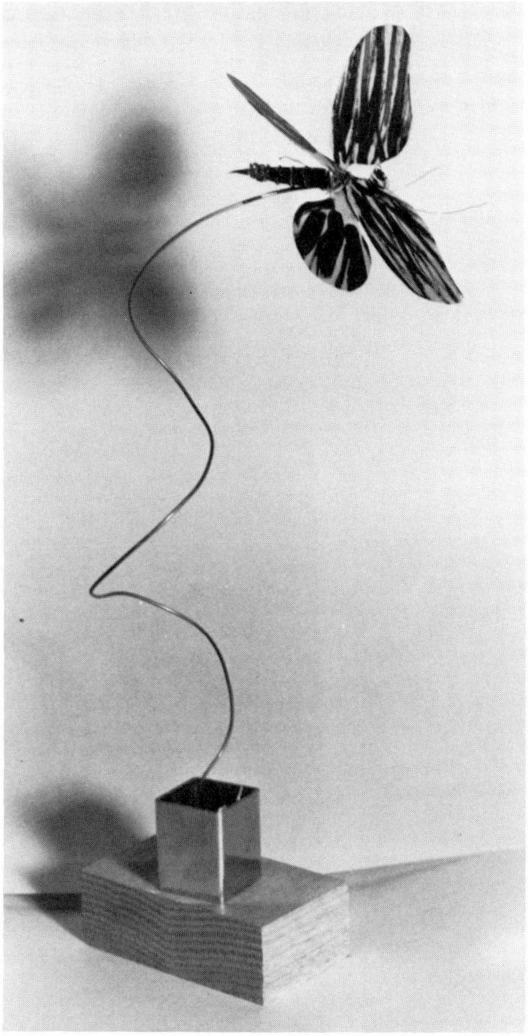

Illus. 37. The wings of this flying butterfly were created with the four-layer method of papier mâché.

The four wings of the flying butterfly in Illus. 37 were created with the four-layer method. A fine wire was placed between layers two and three during the drying process. The wire was allowed to hang out from the wings so that it could be wound round the body of the butterfly, thus holding each wing in place. The butterfly's body consists of a nail with fine wire wound round it. A wood block and scrap piece of chrome provide the base. The end of the wire which was wound round the body was twisted slightly and placed inside the chrome piece.

A salt-box body, wrapping-paper-roll legs, balloon head, and a paper-towel-roll neck are all joined with masking tape to form the foundation for the papier mâché-covered dinosaur in Illus. 38. Tempera decorates the surface.

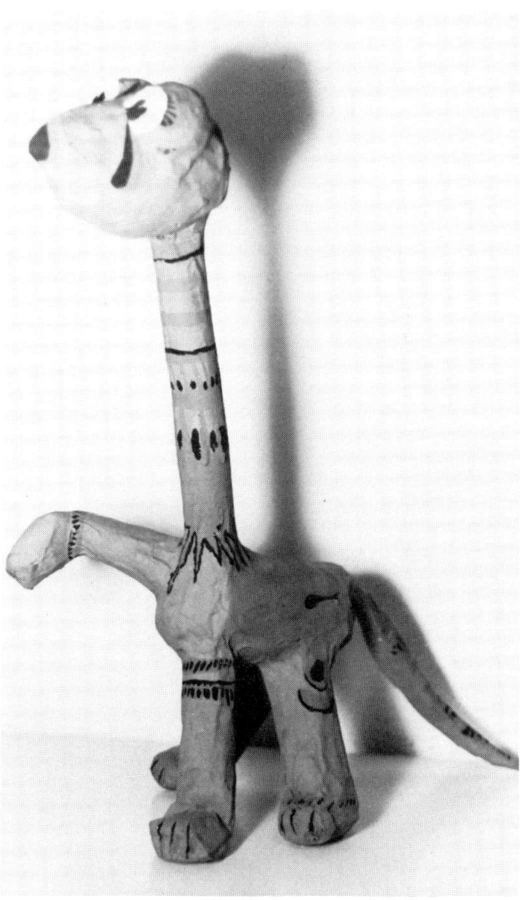

Illus. 38. This dinosaur has a variety of cast-offs underneath his papier-mâché hide.

Illus. 39. This colorful papier-mâché pig has a plastic bleach-bottle body. Pencils, poked into the plastic to serve as legs, were also covered with papier mâché. The surface was decorated with tempera paint.

A colorful papier-mâché bracelet can easily be made by covering an empty tuna-fish can (with both ends removed) with newspaper strips. Cut the strips into narrow widths, run them through paste, and wind them round the can until they are about four thicknesses. Tempera paint can be used for decoration. Shellac or Mod Podge can be added for a protective surface.

The bracelet in Illus. 40 had two layers of papier-mâché strips applied to a tuna can. Then circular plastic-foam packing material and more papier-mâché strips were added. Facial tissue, dipped in glue, were used for the final layers because it adheres very closely to contours, in this instance, the plastic foam.

Illus. 40. A tuna fish can makes an ideal foundation for a papier-mâché bracelet.

Illus. 41. Another tuna-fish-can bracelet is decorated in a different way.

In Illus. 41, cord, placed in an X design, was added to the tuna-fish-can bracelet between the second and third layers of strips. Christmas glitter was added for decoration.

The four-layer method is the best for making flat papier-mâché jewelry as you have seen. The earrings in Illus. 42 were all created in this way. The earrings on the upper left have sequins adorning the tempera-painted faces. Plastic-foam packing material was applied between layers two and three in the large earrings on the right. The three-dimensional effect of the ear-

Illus. 42. These whimsical earrings were fashioned from four-layer papier mâché.

Illus. 43. A unique pendant necklace and matching earrings are decorated with tempera paint, Christmas glitter, and tiny beads.

rings at the bottom was created by applying minute rolls of paper between layers two and three.

Four layers, approximately 2 inches (5 cm.) high by 1½ inches (3.8 cm.) wide were folded once, with the fold located at the bottom of each piece of the necklace in Illus. 43. While it was still damp, wire was inserted through holes pierced with a needle. Three fine wires are twisted together at the top. They were separated and run through the entire height of the piece. Beads were then added. Tempera paint and Christmas glitter decorate the surfaces. The earrings were made the same way.

Illus. 44. When making holes in papier mâché, be certain to do it while it is still damp. You may need an X-Acto knife for large holes, but a needle will suffice for small holes, such as the ones used here for the findings.

Plastic

Illus. 45. This remarkable "Pig with Hair" was fashioned from a plastic bleach bottle. A plastic pipe plug serves as a snout and plastic bottle caps make ideal unpiglike eyes. Cork ears and wood-scrap legs complete the picture. All parts were glued onto the bottle.

Plastic products run a close second to paper in our daily trash disposal. Egg cartons, drinking cups, bottles, containers of various kinds, food wrap, packing material, artificial flowers and fruit, toys, are all among the plastic discards you might have available for recycling. On these pages, you will find many ideas to get you started on the road to plastic recovery.

"Pig with Hair" in Illus. 45 and the "Swedish Girl" in Illus. 46 give you examples of what you can do with plastic bottles, which, of course, come in different shapes and sizes. A large, stout bleach bottle is the ideal choice for the pig, while the girl has a body made of a detergent bottle that is wide at the top, narrow in the middle, and wide at the bottom.

While the pig needed few adornments, the girl offered an opportunity to use up a great number of other kinds of odds and ends of materials on hand. Her apron is made of a scrap from an old white sheet decorated with strips of crepe paper, as is her blouse, which has the addition of construction-paper flowers. Under

Illus. 46. "Swedish Girl" is the title of this charming doll, who has a plastic detergent-bottle body underneath her finery.

the apron is a crepe-paper skirt held in place with an elastic band. Two pipe cleaners were stuck into small holes made in the bottle to serve as arms. An absorbent-cotton-ball head, covered with a piece of flesh-colored nylon stocking, is adorned with yellow-yarn hair and a construction paper cone hat.

Plastic trays and some packing cartons provide an excellent white backing for photomontages. Illus. 47 shows a seasonal theme of winter and spring photos glued to the recesses of two food trays. You might want to use family photographs or photos of your pets or hobbies. Two holes cut in the top and a piece of colorful yarn will allow you to hang your montage.

Plastic packing material used to ship science equipment provides the tesserae for the intriguing mosaic in Illus. 48. The material was cut into many different shapes to provide variety in the design. Black yarn was first glued to a ceiling-tile background and then the tesserae was glued within the shapes created.

Illus. 47. Two plastic food trays form the background for this photomontage.

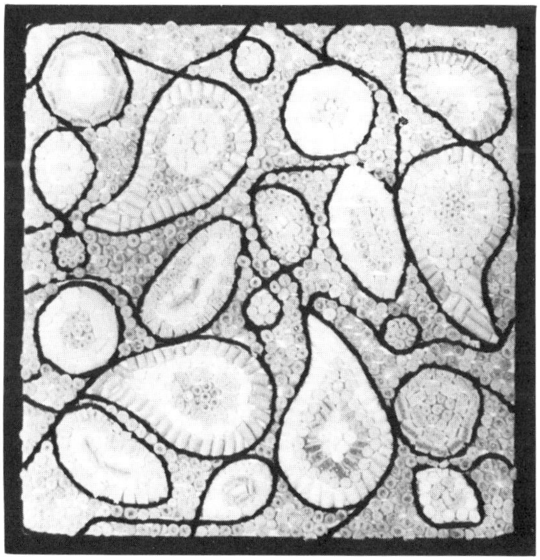

Illus. 48. A mosaic created from plastic packing material is here glued to a background of ceiling tile. Since the material is white, you might want to paint the pieces, either before or after you complete your mosaic.

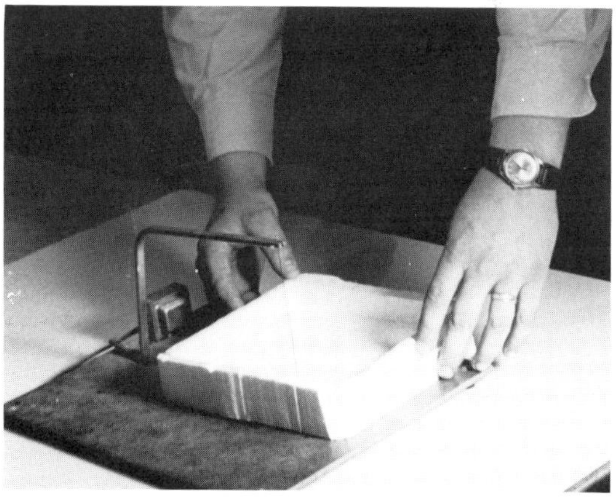

Illus. 49. You can cut plastic foam with a coping saw, but this electric hot-wire cutter melts the plastic quickly and easily and allows you to cut the plastic into whatever shapes you wish.

Illus. 50. This happy clown head is a plastic bleach bottle in disguise! Sprouting from the top of his head is a bunch of plastic flowers. Small pieces of felt decorate his cap and provide his facial features. Old tassels cut from a discarded lamp, ribbon, and sequins provide the final touches to his costume.

The "city" in Illus. 51 is called "Styroville," and is composed of plastic containers glued to each other end to end and side to side, along with a number of paper cups joined together in the same way. The "gate" to the city is made of a piece of plastic foam with the portal cut away by a hot-wire cutter (see page 23). A plastic soldier is charging out from one of the buildings, nearby which stands the top of a plastic lipstick container. All of the pieces are mounted with glue on a triangular piece of Masonite (pressed-board).

The next time you empty a cottage cheese or whipped butter container, why not make a hanging planter like the one in Illus. 52? An assortment of small seashells collected on an excursion to the shore at some time and then set

Illus. 51. "Styroville" is made up of a variety of plastic cups and containers.

Illus. 52.

aside proved to be the ideal decoration for this pot as you can see. After the shells were glued to the container and were thoroughly dry, more glue was spread evenly among the spaces between the shells. While the glue was wet, sand was sprinkled on it until it was completely covered. You will have to do this in sections, so that you can lay the container down on a dry side while you are sprinkling.

If you don't have shells handy, you can use almost any small decorative objects, or even an assortment of different kinds, such as buttons, beads, or whatever. If you want, you might suspend the planter with a macramé hanging, or simply use heavy string, and knot it as shown.

The stunning room divider in Illus. 53 shows you what you can do with used plastic foam cups! Immediately after use, wash them out, set aside, and when you have a huge collection, try making this easy-to-do divider. Glue the cups top to top, and bottom to bottom, row by row until you have reached the size and number of rows you need to fill whatever space you have. Then glue the rows together as shown here. Since the divider is very light in weight, you can suspend it from the ceiling or in a doorway easily. Finish it off by spraying with a rich gold or brownish paint, which will help to give it an illusion of greater solidity.

Illus. 53.

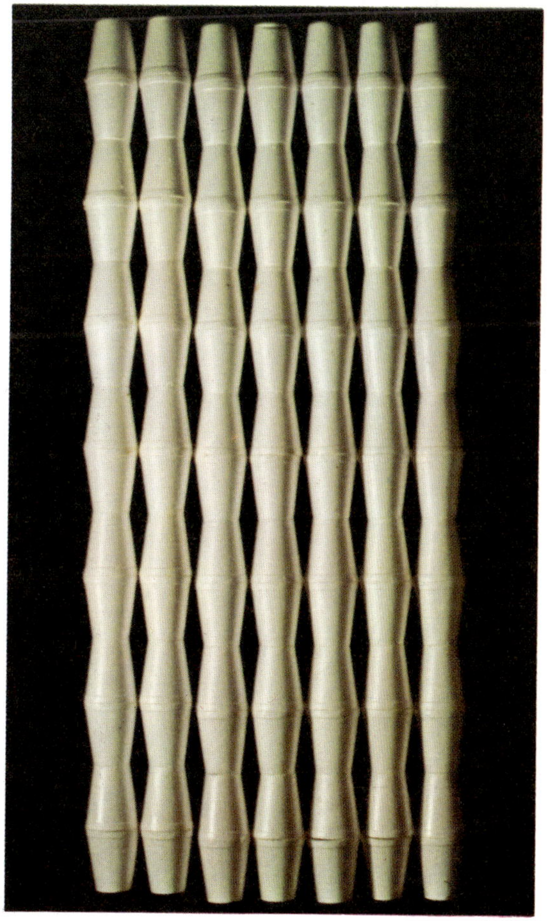

Illus. 54. Don't throw away those plastic aerosol can tops! Make a striking bas relief such as this one entitled "Stop and Be Careful."

The bas relief in Illus. 54 shows what you can do with plastic aerosol can tops. The background is a $\frac{3}{4}$-inch (1.9-cm.)-thick piece of plywood measuring 18 × 30 inches (45.7 × 76 cm.), painted with black tempera. Before the other materials were added, plastic spray was applied to the painted surface. The colorful tops were glued to the wood with plastic glue. Plain white drawing paper, cut into different shapes, fills in some of the spaces between the tops and also serves as circular bases for the dark-colored tops.

Illus. 55. Egg-shaped plastic containers for panty hose are here cut in half, painted, and glued to a sheet of painted plywood to form another striking bas relief.

Illus. 56. Any music lover would enjoy having this decorative guitar hanging in his room. Cut from a flat sheet of plastic foam with a hot-wire cutter (see page 23), the guitar was then sprayed with silver paint, and yarn "strings" pinned on.

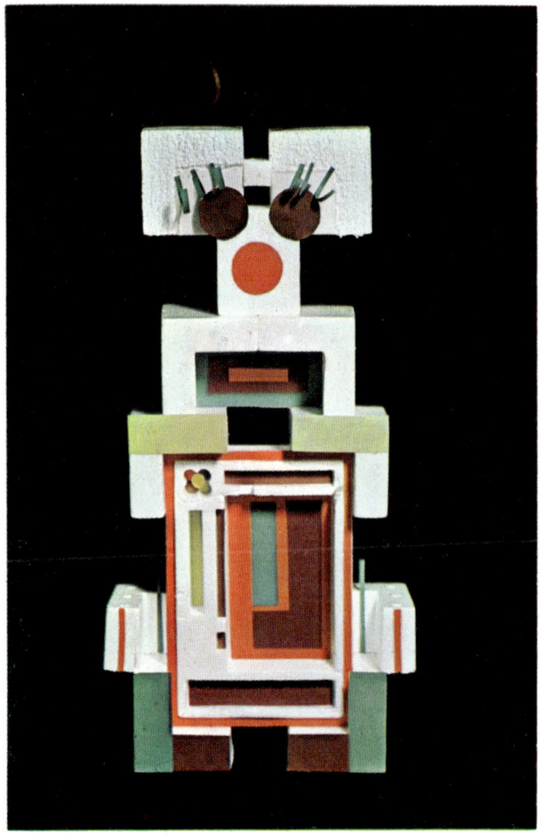

Illus. 57. Either cut pieces of plastic foam or save all the odd shapes you find used as packing material for breakables and turn them into a robot. Construction paper adds touches of color and glue holds everything together.

Illus. 58. Cut plastic foam is here glued together and sprayed with silver paint to form a fantasy castle.

Illus. 59. Here's an idea for a clever bird feeder for your back yard.

Bird lovers might want to make the unique bird feeder in Illus. 59. Cut a 4 × 5-inch (9 × 12.7-cm.) hole in a gallon size plastic bleach container as shown in the illustration. Bend the resulting flap up to serve as an awning. To decorate the feeder, apply small, unevenly cut pieces of masking tape all over the bottle. Here, horizontal strips were applied across the lower part of the container to suggest bricks, and vertical strips on the upper part to suggest shingles. Use a black or red soft-tip ink marker to outline the edges of the pieces of tape to emphasize the design.

If you want, you can add brown shoe polish to the tape to deepen the brick and shingle tones. Finally, make a small hole in the top of the bottle and insert a section of a coat hanger and your bird feeder is ready to receive guests.

Plastic six-pack soda rings can take on a new look in an unusual tote bag. Place a group of rings on a sheet of aluminum foil, overlapping them as shown in Illus. 60. Then place a second sheet of foil over the plastic rings and run a hot flat iron over it to fuse the rings together. Do this with two separate designs of the same size and then sew the two pieces together using a simple blanket stitch. Here, buttons and window-shade rings were sewn to the plastic for decoration. For a handle, you might braid together approximately 30, 20-inch (50-cm.)-long pieces of yarn and attach them with thread to the plastic rings. Naturally, you can make your tote bag as big or small as you wish, depending upon how many plastic rings you have saved up.

Illus. 60. Save your plastic six-pack soda rings and transform them into a decorative tote bag!

You may not have realized that the fascinating "thing" in Illus. 61 was a lamp when you first looked at it, but that's what it is! The base is a green plastic electrical coffee pot and the shade is composed of a number of green plastic berry containers.

To begin, remove the glass knob in the middle of the percolator top and insert the electrical cord, enclosed in a metal tube which will act as a support for the bulb and socket above. Wire the lamp in such a way as to make use of the electrical rigging in the percolator. That way, you simply plug in the coffee pot's cord, and your lamp goes on.

Illus. 62 shows the procedure for glueing the berry boxes together for the lamp shade. In Drawing A four boxes were stacked and glued together. You will have to make four stacks. Notice in Illus. 61 that the bottom box of each stack extends out beyond the upper three boxes. Drawing B shows the lamp shade as seen from the top, that is, looking down into it, after the stacks were glued together. The arrows indicate where sheets of plastic packing material with a bubbly texture were glued to the boxes to diffuse the light of the bulb.

Additional berry boxes, open-side up, were glued here between boxes 1 and 2, 2 and 4, 4 and 3, and 3 and 1 (Drawing B). Four more boxes were placed on an angle inside the top boxes. Soda-can tabs painted green, plastic grapes, and yellow rick-rack decorate the completed lamp.

Illus. 61. If you need an extra lamp, look what you can do with a plastic coffee pot and some berry containers.

A

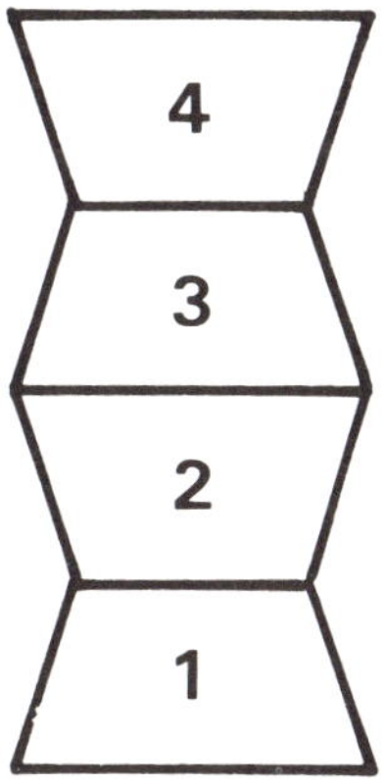

Illus. 62.

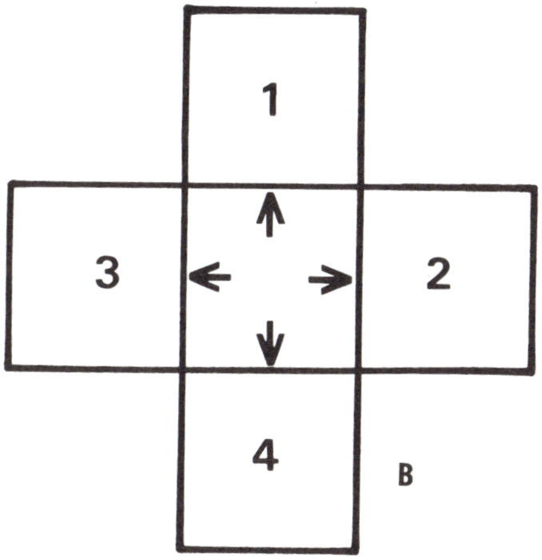

B

Natural Materials

Illus. 63. "Tree Trunks" makes use of leftover coffee instead of paint!

We could all learn much from Nature—probably the greatest recycler of its own "waste" in the world. However, although most cast-off natural materials do serve a purpose in the cycle of Nature, we can always lend a helping hand, as this section will prove to you.

For example, did you throw out the remains of the morning coffee today? If you did, you missed the opportunity to make a "coffee painting"! Illus. 63 and 64 show two different techniques. Illus. 63 is a traditional semi-abstract painting using coffee as the medium. By varying the water content, different gradations of browns were easily achieved. In Illus. 64,

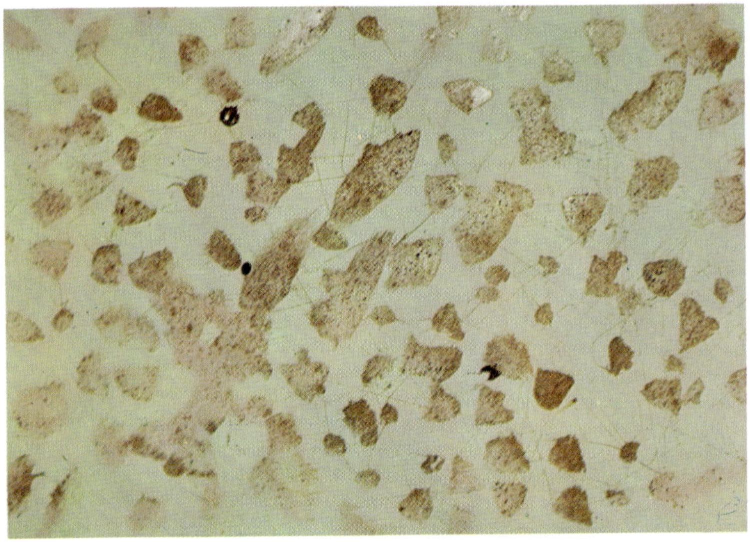

Illus. 64. Another coffee painting, but with a difference. Here, coffee is dribbled onto rubber cement to create a charming ethereal effect.

rubber cement was dribbled onto white paper and then dry instant coffee was crumbled on the wet cement, creating interesting shapes and texture.

Illus. 65 gives you a good idea of what you can do with all those interesting bits of bark, moss, branches, and so on, that you might have picked up in the woods and set aside. A cork bulletin board forms the background for this nature collage, but of course, you can use any sturdy background material you have handy. Pieces of bark from birch, oak, cedar and moss-covered hardwood trees are here glued to the board in a pleasing pattern. Tacks and pins are helpful in holding the objects in place during the drying process. If necessary, you can leave some of the tacks in if your objects are too heavy to stay put with glue.

Illus. 65.

Illus. 66.

A piece of driftwood provides the background for the wall hanging in Illus. 66. The theme of the collage is based on a selection from the Bible, "To everything there is a season and a time to every purpose under heaven," which is written on a piece of paper whose edges were burned to create an antique effect. A bird's nest, dried weeds, peach pits, nuts, a seashell, a twig, a cat-tail (Timothy grass), and a pine cone are the basic ingredients here. The small turtles and frogs are made of peach pits, assorted nuts, and pipe cleaners. The eggs in the nest are pecan nuts colored with a soft-tip ink marker to simulate robins' eggs.

The striking wall plaques in Illus. 67 were created with Osage oranges, which are very large and have a bumpy skin. Osage oranges are inedible and are not found everywhere, so you might use over-ripe oranges, tangerines, or grapefruit for your collage. First, cut the fruit into thin slices. Then cut slits into the edge of each section to simulate flower petals. Lay the sections on a cookie tin and bake them in the oven at 250°F. (93°C.) for two hours. The longer you leave the fruit in the oven the darker it will become. Cool, lay the sections out on a stiff background in any arrangement that suits you, and then glue them in place.

Illus. 67.

Illus. 68. Cat-tails, grasses, shells, a starfish, feathers, sand, and bark are here glued to a construction-paper background, which in turn is glued to stiff cardboard. A border of fish netting completes this marine collage.

Illus. 69. Here, large pieces of driftwood were glued to a flat board first, and then smaller pieces of driftwood and a number of different kinds of shells were used to fill in the spaces and thus cover the board. White glue was used throughout.

Illus. 70. For a change of pace, why not make a nature diorama? Here a cardboard box is lined with light blue construction paper to suggest the sky. Dried flowers, taped together into bunches resembling trees, are glued to the paper background. Leaves are scattered round the floor of the diorama and chalk shavings are sprinkled over the entire scene to suggest falling snow. (To make shavings, run a scissors blade over a piece of chalk, or if you want, use left-over Christmas snow.)

Illus. 71. Dried flowers, moss, bark, and milkweed (milk parsley) pods arranged on a piece of barnwood create this setting which almost perfectly simulates a scene in Nature.

Illus. 72. If you have a piece of scrap wood you don't know what to do with, here is one idea. This piece of wood was painted light yellow and then covered with a layer of Decopage. Torn pieces of orange, yellow and brown tissue paper were then placed on the wood and also covered with a coat of Decopage. Then another layer of Decopage was painted over both the wood and the tissue. Odds and ends of driftwood were assembled to form a tree and dried mushrooms were placed at its foot. Decopage also gives these a glossy effect.

Illus. 73 (left) and 74 (right). Pieces of straw, softened in water, were split open, flattened and cut into pieces to form these two designs on felt backgrounds. Be sure to dry straw thoroughly before glueing it.

Illus. 75. Plywood, painted with a mixture of India ink and shellac, provides the background for an interesting array of sumac, evergreen twigs, wild berries and foxtail weeds. Transparent tape and tacks hold the materials to the wood.

Illus. 76. If you have a lamp or a vase or anything similar that needs a new face, clean out your cupboard and mask its faults with a mosaic design.

Nature provides an abundance of materials, large and small, which are suitable for making mosaics. If you have small quantities of such edibles as rice and macaroni left in the bottoms of boxes, and a lamp that needs a little rejuvenation, you can turn out a lamp such as the one in Illus. 76.

Rice kernels were soaked overnight in dark green and red food coloring and some round macaroni was painted with dark blue acrylic paint. Horizontal borders of blue macaroni were glued to the base about an inch (2.5 cm.) apart, and then alternating colors of red and green rice were used to fill the spaces between.

Some other ideas for your nature mosaics are shown below and on the next two pages in color.

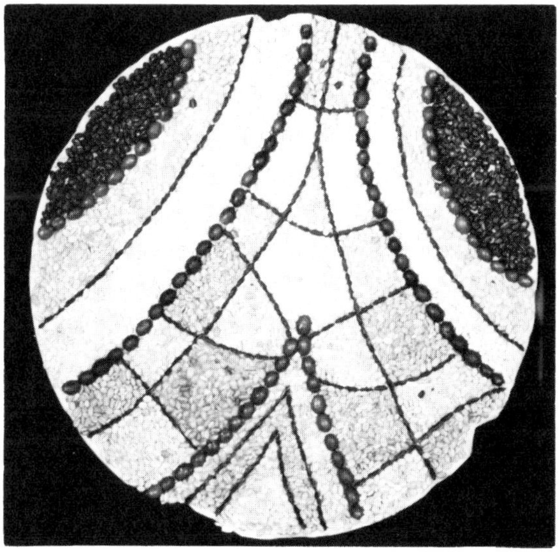

Illus. 77. A china plate forms the background for this circular mosaic. Hot-water fabric dyes were used to color the materials and the resulting orange, yellow, blue and white barley, green rice, and maroon and blue beans were glued to the plate. A short soaking of five minutes is sufficient to color these materials.

Illus. 78. Macaroni of different kinds—round, shell, and spiral—was painted with tempera paints and glued to a round platter, along with orange and yellow aquarium gravel.

Illus. 79 (left). This colorful floral arrangement is composed exclusively from rice dipped into food colorings. The plywood background was covered with glue and then the rice was applied with a tweezers. Illus. 80 (right). Here, rug filler (yarn) was glued upon a plywood board to outline the design. Then aquarium stones of various colors were glued within the outlined shapes.

Illus. 81.

The four sides of the sleek looking Parsons table in Illus. 82 were sawed from $\frac{1}{2}$-inch (1.27-cm.)-thick plywood. The openings measure 12 × 12 inches (30 × 30 cm.). The mosaic top, shown in Illus. 81, consists of stones attached with epoxy cement to a piece of $\frac{1}{8}$-inch (.3-cm.)-thick plywood, cut to fit within the four sides. A 1 × 1-inch (2.5 × 2.5-cm.) ledge was nailed inside the top rim of the table to hold the stone mosaic. The ledge is approximately 1 inch (2.5 cm.) from the top rim.

When the epoxy cement was dry, white tile grout was spread over and between the stones. The excess grout was then removed cleanly to expose the stones. A temporary wooden frame

Illus. 82. This plywood Parsons table has a stone-mosaic top.

was built round the mosaic to hold the grout in place. You must be sure to remove the excess as quickly as possible before it dries and hardens. To protect the mosaic from stains, place a piece of glass or acrylic sheet over it. The mosaic inset measures $15\frac{1}{2} \times 15\frac{1}{2}$ inches (39×39 cm.).

To complement your table, why not make a table-top sculpture such as the one in Illus. 83? This unique fisherman is composed completely of stones, all glued together with epoxy cement.

The base stone that the fisherman is sitting on was first glued to a small concrete slab. Then, in stages, the torso, head, face, arms, and legs were assembled in that order. For each stage of development, the glue was allowed to dry overnight. This, of course, means that your sculpture will probably take a few days to complete. Some of the stones had to be propped up in place during the drying process. Some leaned against old boxes, others were supported by crushed paper, and some were even mask-taped in place during the drying. The little fish was put together in the same manner. A stick fishing pole with a string was added last.

Illus. 83. Here is an all-stone sculpture you might use to decorate your stone-mosaic table.

Illus. 84 (left). "Nesting Eagle on a Love Tree" is the descriptive title of this nature assemblage. The tree is composed entirely of pieces of found driftwood, glued and nailed to a driftwood base. Small holes were drilled into the tree and then toothpicks were inserted into the holes to act as "pegs" for attaching limbs, mushrooms, and "animals." Illus. 85 (right). This striking stabile is also composed of driftwood. All of the pieces are held together with nails and screws. Be sure to pre-drill holes to avoid splitting the dry wood.

Illus. 86. This table-top stabile is made up of bark, fungus, pine cones, dried insects and driftwood. White glue holds everything together.

Next Christmas, you won't have to buy a table tree if you make the nut tree shown in Illus. 88! (An added bonus is eating the nuts at Christmas dinner.) You will need a plastic foam cone, assorted nuts, pine cones, red tree ornaments, round toothpicks, a hammer, one nail and green plastic leaves.

With a hammer, make a hole in each nut by lightly pounding a nail into it. Then insert a toothpick into the nut as deeply as possible. Insert the other end of the toothpick into the plastic foam. You can glue the other ornamental objects into place. If you don't have a plastic foam cone available, you can cut a piece of tagboard into a circle, make a slit from the edge to the exact center (use a compass). Then, overlap the resulting edges until you have a perfect cone shape. You can either poke the nuts and other decorations into the tagboard as described above, or glue them in place with fast-drying airplane glue.

Illus. 87. To make a festive nut tree for Christmas, you need little more than nuts, a plastic foam cone, and some toothpicks.

Illus. 88. Here is the completed nut tree, all ready to eat. You can add any small ornamental objects that you wish to spruce up this holiday tree.

Illus. 89. This striking piece of beach sculpture is constructed on a 5 ft. × 11 in. (1.5 m. × 28 cm.) board. The pieces of driftwood are held to the board by wood screws inserted into holes drilled in the board from the bottom side. Walnut stain was used to darken some of the wood pieces. Other materials included in the sculpture are rope, fish netting, stones, sea shells and coral. These smaller and lighter materials are glued onto the board.

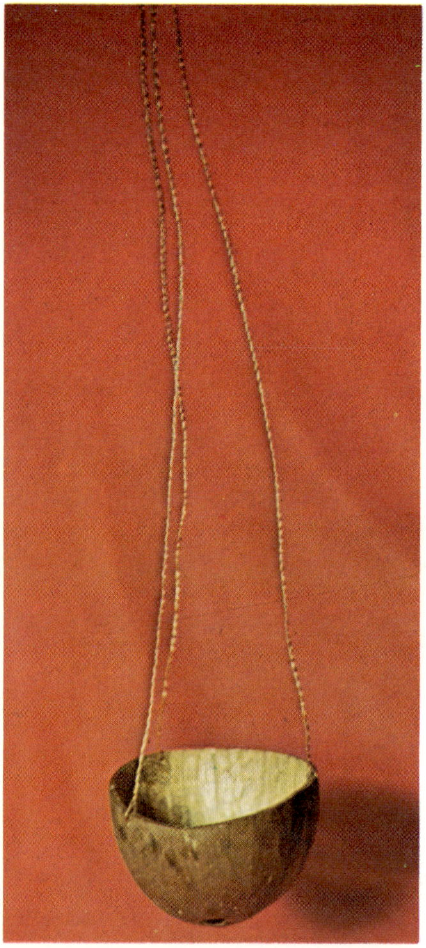

Illus. 90. A coconut shell, sawed in half, makes an effective hanging planter. Three holes, drilled near the top rim, allow cord to be passed through and knotted.

Illus. 91. An idea for all those small shells you have collected is to make a necklace with them. Use a needle or a nail to make the holes for stringing and smooth them off with soft sandpaper. However, since the shells have a tendency to break easily, a fine drill would be preferable. Here, a leather strap was used for stringing.

Some ideas for nature mobiles are shown in Illus. 92, 93, and 94. All of the hanging and cross pieces are made of driftwood. Screw eyes (the kind used for hanging pictures) were inserted into each piece and fish line was run through each eye to suspend the mobiles.

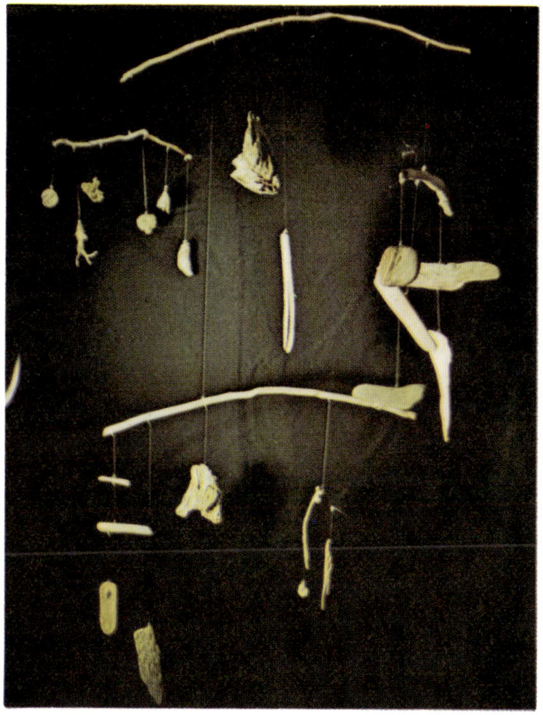

Illus. 92.

Illus. 94.

Illus. 93.

Illus. 95. A mobile which consists of objects collected on a honeymoon is made by attaching the objects with fishing line.

Illus. 96. Large pine cones form the basic material for this mobile. Wire, cut from coat hangers, form the cross pieces and fish line holds each hanging piece.

In the mobile shown in Illus. 95, small holes were drilled (with a hand drill) and fish line was inserted into the holes to hold the hanging pieces. The line was first tied to the wood and shell pieces and balance was established by switching the pieces around. Then the line was threaded and knotted to make the mobile a permanent construction. Entitled "Honeymoon Mobile," it consists of objects collected by a young couple on their honeymoon—driftwood, dry cat-tails, seashells, and feathers.

Another interesting and very different technique you might try is sand painting. To begin, add one tablespoon of dry powdered tempera to three cups of beach sand. Very little tempera is needed to color sand, making this a very inexpensive venture. Draw your design on

Illus. 97. You began working with natural materials in this section by making a coffee painting. Now you can try another painting technique—sand painting.

Illus. 98. The sand painting begun in Illus. 97 is here complete with dark yarn outlining the design elements.

tagboard, cardboard, or Masonite (pressed-board). Then apply white glue within the shapes you have created and drop the colored sand on the shapes as shown in Illus. 97. When all the colors have been dropped you can glue yarn round the parts of the design to emphasize it, as shown in Illus. 98.

Illus. 99. Here, yarn was not needed to heighten the contrast of the various segments of this brightly colored sand painting, which is mounted on a piece of ceiling tile. Rick-rack is glued on to form a border for the tile.

Illus. 100. A plastic kitchen fork, pressed into red ink and printed upon yellow tissue paper, creates a rhythmic effect in this gadget print.

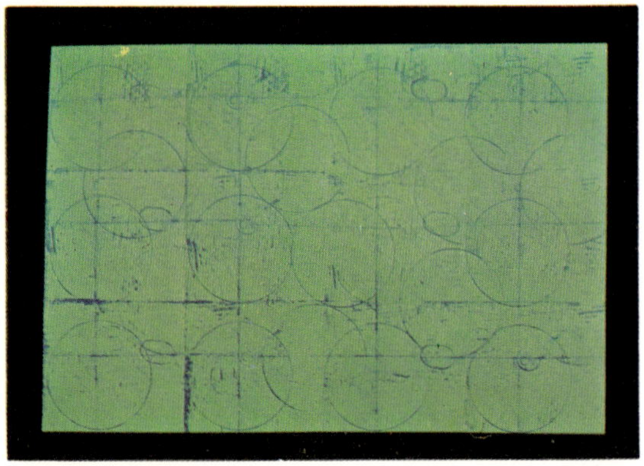

Illus. 101. Here, the rim of a tin can was dipped into ink and pressed onto construction paper to create the delicate circular design. A small, oval-shaped pillbox provides the second, smaller form, and the edge of a sheet of cardboard was used to create the straight lines.

Printing with Scrap

Without knowing it, you probably have a treasure trove of printing materials right in your kitchen, closet, basement, desk drawer, or wherever you squirrel away odds and ends of things for a rainy day. There are three requirements for printing. First, you need some kind of shape with which to print. Then you need a printing surface such as paper, cardboard, wood, newspaper, tissue paper, or cloth. Last, and the only item you might need to purchase, is ink. You can use either water-soluble or oil-base printer's ink. However, as a substitute, mix tempera paint with liquid starch, or use acrylic paints just as they are.

The wide variety of printing gadgets used here will give you many ideas to use as springboards for your own creative printing from scrap.

Spread your ink or paint on a glass surface using a brayer (rubber roller). When the ink sounds tacky (sticky), press your gadget on the inked glass and then press it on the printing surface. (You might use pliers to hold small articles such as keys.) As an alternative, brush the ink on your gadget, rather than on the glass if you prefer.

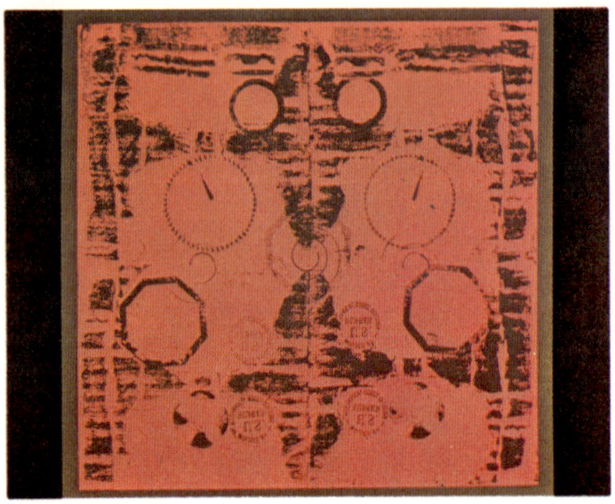

Illus. 102. A rubber stamp, a clock face, an empty face-powder box and pieces of ribbed cardboard produced the symmetrical design in black ink on red construction paper.

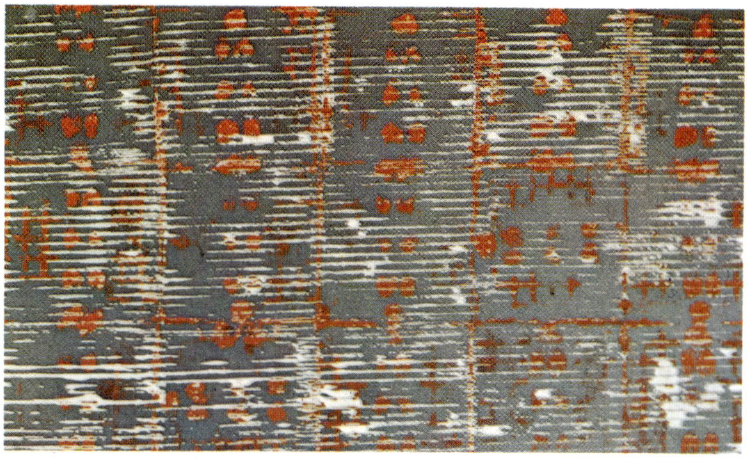

Illus. 103. A plastic vegetable container produced the small red shapes of this design, and a rubber mat was used to create the light, horizontal, closely spaced lines. The background is charcoal construction paper.

Illus. 104. Orange construction paper forms the basis for a unique print made using an old latch key, a kitchen fork, and the end of a wood block.

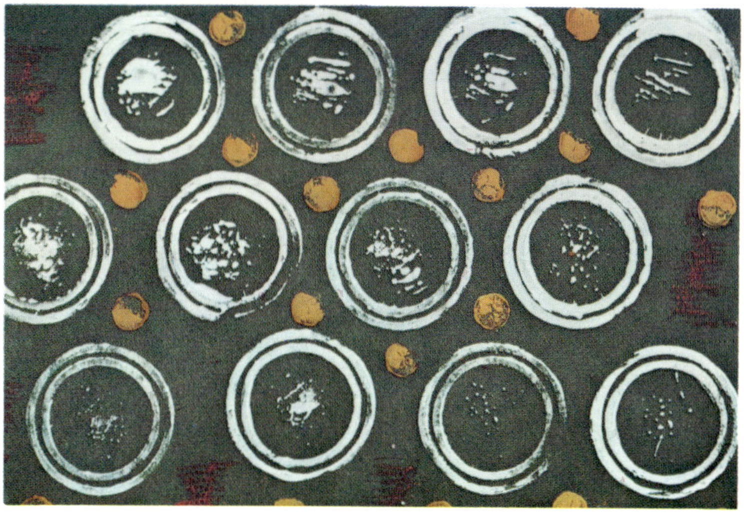

Illus. 105. This rhythmic design was achieved with a jar cap, a corn skewer, and the head of a bolt.

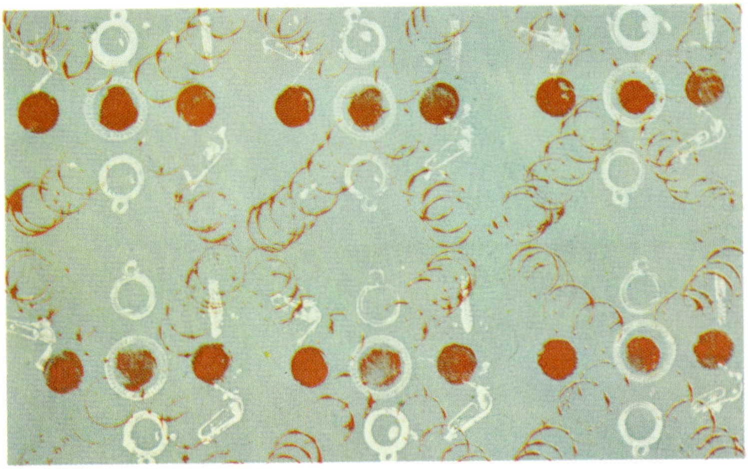

Illus. 106. Aqua-colored construction paper is a pleasing background here for a print created with a poker chip, a bottle cap, and a yo-yo ring.

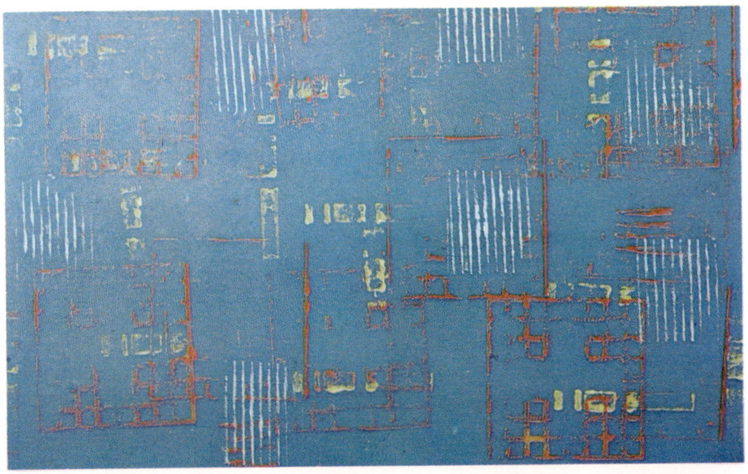

Illus. 107. A plastic mat, half of a clip-type clothespin, and a plastic strawberry container here produce interesting lines and shapes in another gadget print.

Illus. 108. The flowerlike effect in this print is the result of using the end of a hair roller. Other objects used include a metal spring, a fork, a metallic button and a fingertip.

Illus. 109. A fork, a jar lid and a faucet combined forces here.

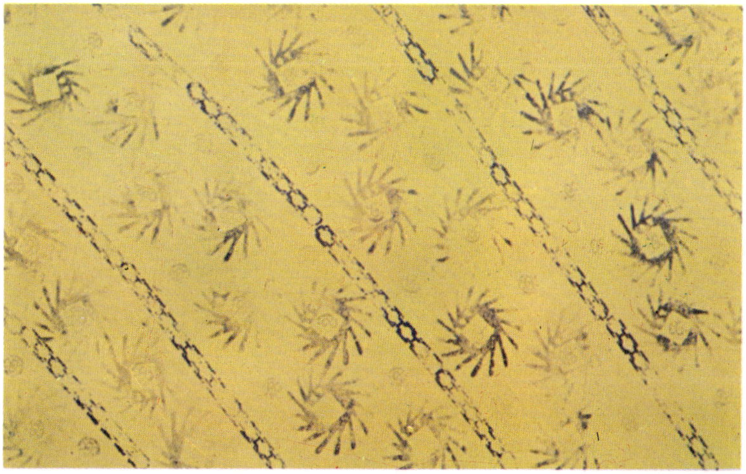

Illus. 110. This diagonal pattern is the result of printing with a meat grinder attachment and the end of a screw.

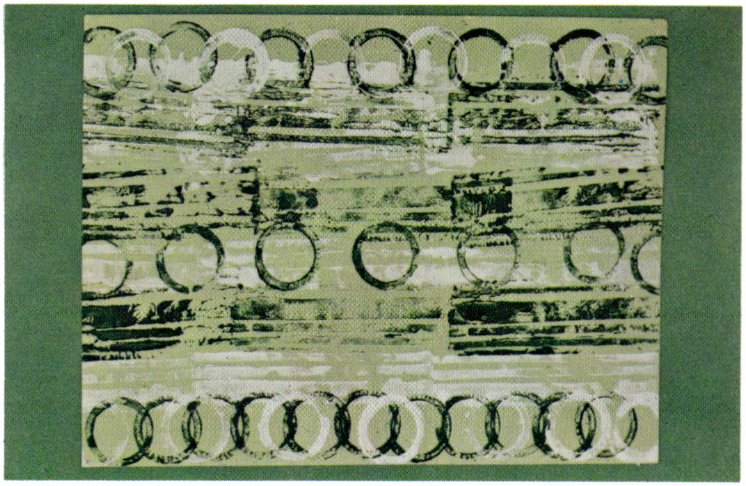

Illus. 111. A block of wood with strips of tagboard glued to it achieved the white and green lines in this design on light green construction paper. A plastic bottle cap made the circles.

Illus. 112. A rubber-soled boot was responsible for the interesting pattern here.

Illus. 113 (top right). A part of a door lock created the red shapes, and a soap holder the blue printing in this design. The background is pink construction paper.

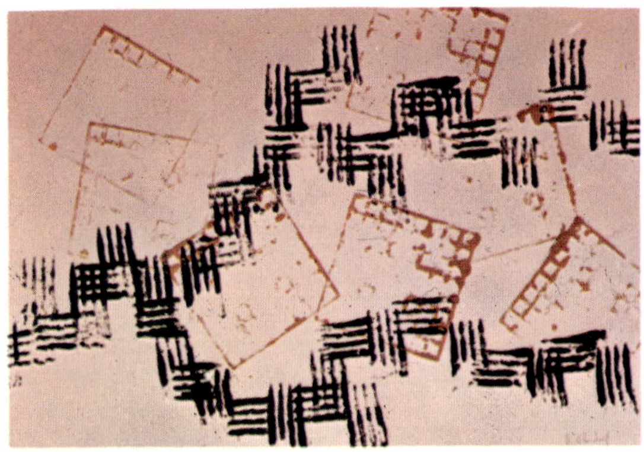

Illus. 114 (middle right). Again pink construction paper forms the background, but this time the printing shapes were a plastic strawberry box and a kitchen fork, producing the black and red designs.

Illus. 115 (bottom right). When you want a design of this type, it is helpful if you crease the paper lightly before beginning so that you can use the resulting lines as guidelines. Here, brown construction paper is the basis for the printing of a sewing-machine bobbin and a car key in orange and white.

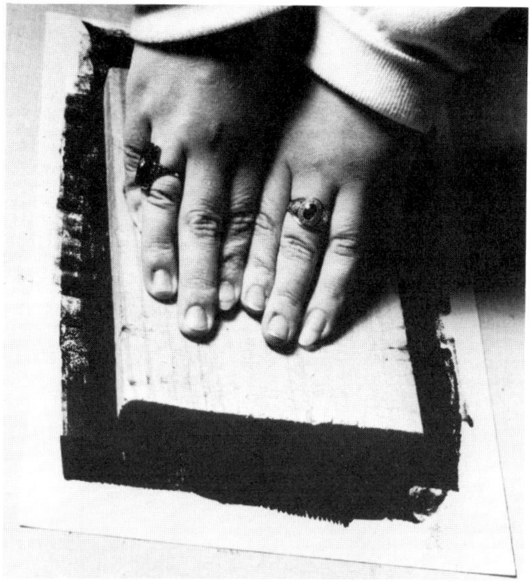

Illus. 116 (left). To print with cardboard, cut up scrap pieces into the various elements of your design as shown here. Then glue the pieces onto a wooden board. If you plan to make many prints, it would be wise to paint the board with shellac to make it waterproof (or in this case, inkproof). Then (Illus. 117, right), ink the glass and press the cardboard down on it, making sure that all parts of the design are inked evenly. Remove and then press onto your printing surface.

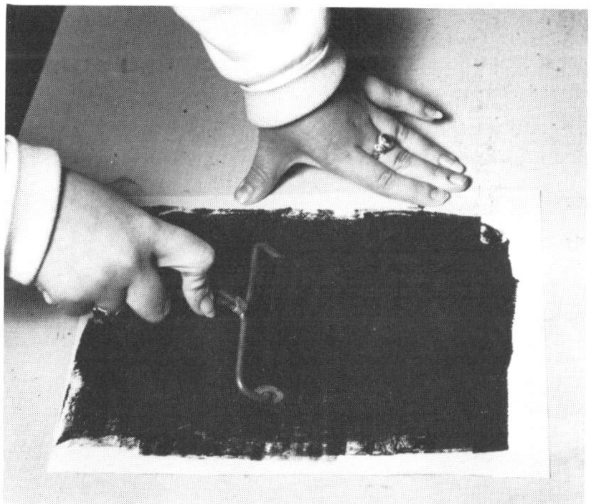

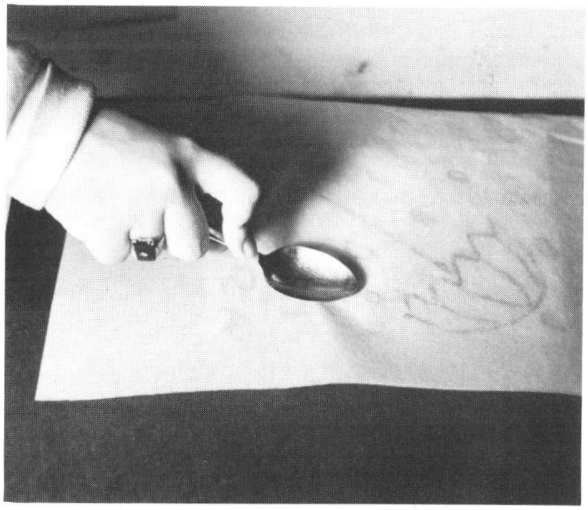

Illus. 118 (left). If you prefer, you can run a brayer (rubber roller) over the inked glass to pick up the ink. Then run the brayer over the cardboard design on the piece of wood, and proceed with your printing on the paper. Illus. 119 (right). You will have to exert some kind of pressure to achieve a distinct image. For example, here, the bottom side of a large spoon is being used to press tissue paper against the inked cardboard design on the wood block.

Illus. 120. Here, an interesting cardboard shape was glued to a wood block, inked, and pressed on pink construction paper.

Illus. 121. This church print uses both the positive and negative sections of the cardboard design. The church was cut from a single piece of cardboard, and glued to a block as was the piece of cardboard remaining after the church was cut out. Then, both were inked and pressed alternately on the background to form this print.

Illus. 122. Ribbed packing cardboard produced this flock of dark green birds on white paper.

Illus. 123. Here, ribbed cardboard is used again to create a totally different design, giving you an idea of the possibilities of printing with this common scrap material. Yellow and red inks were used on a white drawing-paper background.

Illus. 124. Pieces of scrap fabric are ideal to use for your printing. However, thick, or plush fabrics are better suited than fine, thin ones. You can use them in the same way as cardboard. Felt scraps, glued to a wood block, printed this striking example on purple construction paper.

Illus. 125. Pieces of velvet glued to a wood block provided an excellent printing medium for this swirling composition. When glueing a material that has "give" such as velvet to the wood, be sure it is as taut as you can make it. Some possibilities for your fabric prints are carpet remnants, upholstery fabric, burlap (hessian), towels, porous dish cloths, and corduroy.

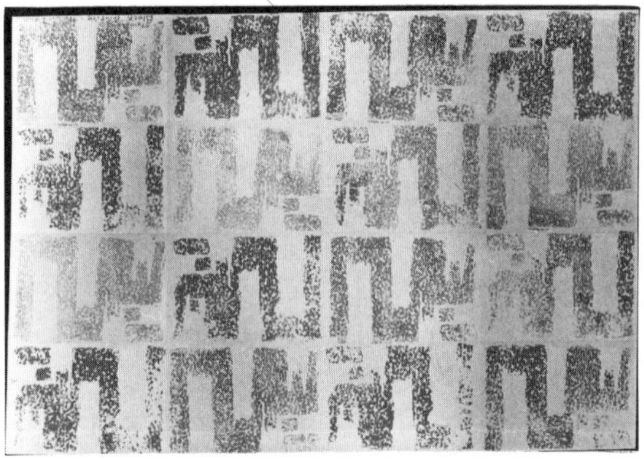

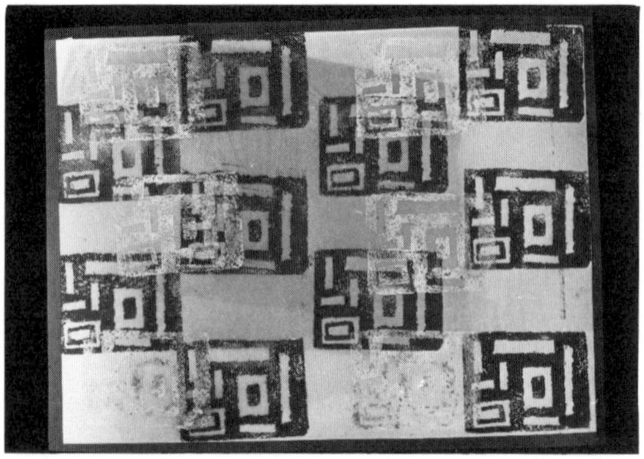

Illus. 126 (left) and 127 (right). Plastic foam, which is often used as a packing material, is a very effective printing medium. Simply use a ball-point pen to impress designs into the plastic. Then ink it, and print. You can also use plastic foam drinking cups and containers. In the example on the left, negative portions of the design were cut away from a thick piece of plastic foam with a pocket-knife. This print was made on yellow construction paper, while the one at the right was made on tissue paper.

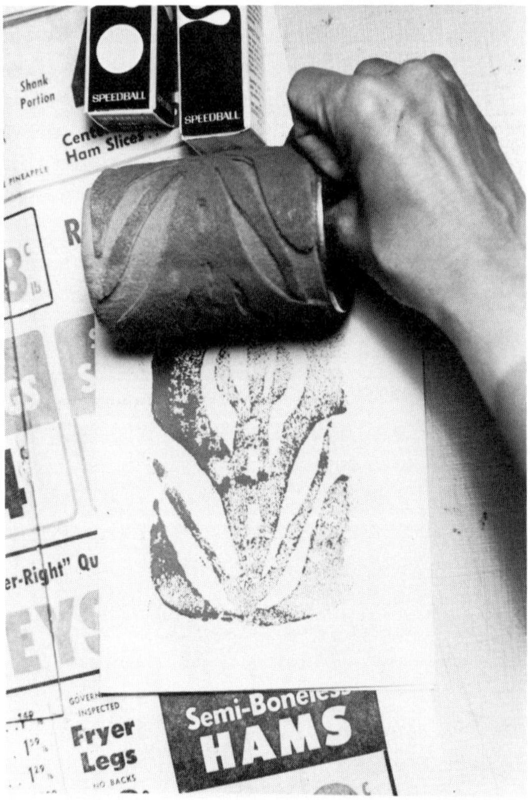

Illus. 128 through 132 are samples of the possibilities of tin-can printing. Remove the label and then both ends of a can, so that you can put your hands inside when rolling it during the printing. Since the surface of the can is slippery, it is often difficult to glue scrap materials to it, although it is possible (Illus. 131). The simplest solution is to glue either a piece of cloth or paper round the can to provide a better surface for glueing.

Be sure the scrap material design is of an equal thickness all over the can so that the entire print will be distinct. When inking the can, begin with the seam-side down against the inked glass, and roll, exerting an even pressure, until you come back to the seam again. Make your print in exactly the same way, starting at the seam. Re-ink and continue with your next print. This is a good method for printing gift-wrap papers and greeting cards.

Illus. 128. In this tin-can printing, the can is covered with a solid piece of felt, on which a design made of felt scraps is glued.

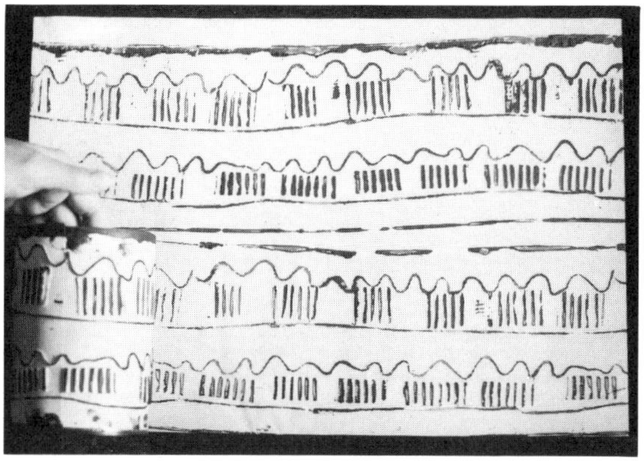

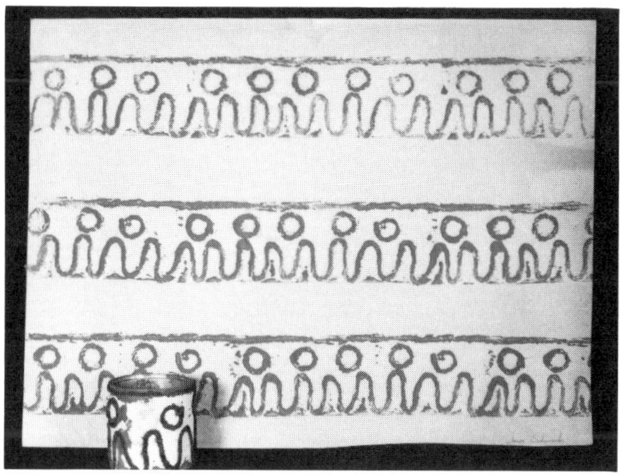

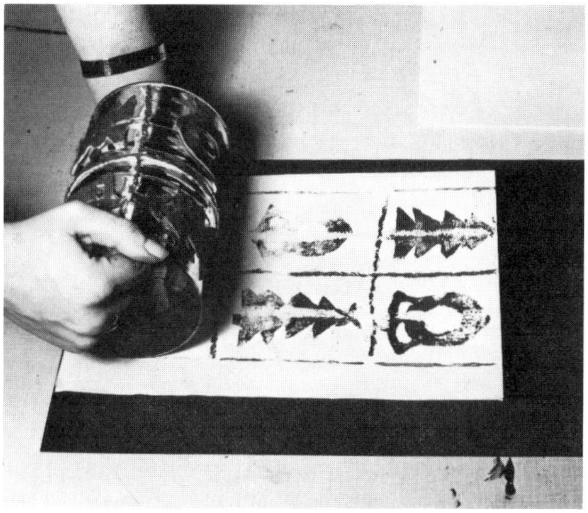

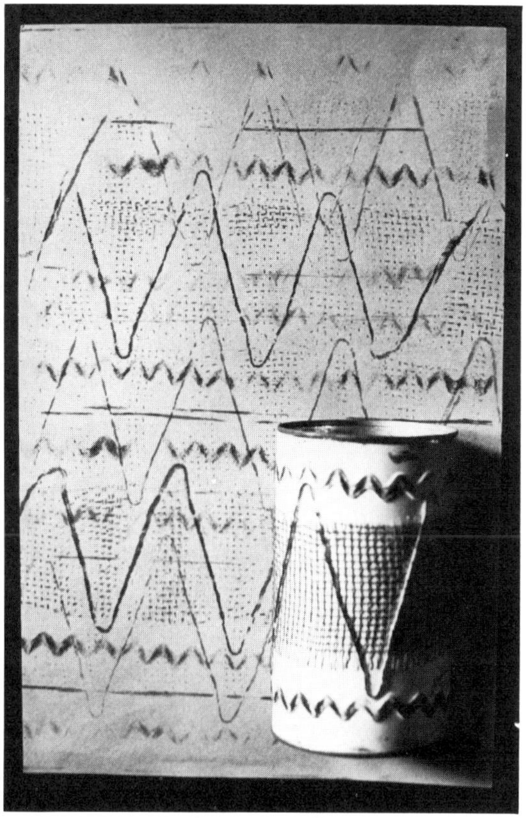

Illus. 129 (top left). Pieces of ribbed cardboard and string are here glued to a tagboard backing surrounding the can.

Illus. 130 (middle left). Heavy yarn, glued onto tagboard in two different designs—a wavy line and small circles—produced this print.

Illus. 131 (bottom left). In this case, a cardboard design was glued directly onto the metal can to create a Christmas card.

Illus. 132 (above right). A section of a mesh vegetable bag, string, and rick-rack form this interesting pattern.

Illus. 133 (left). This is one of Marjorie Koster's woodblocks, which is a work of art in itself. On the left you can see such scrap materials as pieces of metal, plastic doilies, mesh bags and cardboard. The resulting print is shown in Illus. 134 (right).

The prints and woodblocks on this page were made by Marjorie Koster, a printmaker who works on a large scale, the blocks and prints measuring over 3×4 feet ($1 \times 1\frac{1}{3}$ m.). Using woodblock printing tools, she often gouges out words from the block and then glues, nails, or screws a great variety of scrap materials onto the block. She brushes a milk-base glue over the entire surface and allows it to dry before inking the block with lithographic oil-base inks, using a brayer. Then, a large sheet of paper is placed over the block, and Marjorie rubs the back of the paper with an assortment of wooden dresser drawer knobs and kitchen cabinet knobs. After a thorough rubbing, she pulls off the paper and allows the print to dry.

Illus. 135 (left). A variety of metal disc shapes removed from old machinery of different kinds provide the materials screwed, nailed, and glued onto the woodblock to produce the exciting print in Illus. 136 (right).

Fabric

Illus. 137. A circus theme is here rendered with pieces of scrap felt and cork on a canvas background.

There are many sources for your scrap fabric creations. For example, if you know anyone who sews, they might be happy to give you their inevitable pieces of left-over cloth. In addition, worn garments, and cloth bags, such as burlap or canvas, can all be washed and used.

Many of the suggested projects shown here involve appliqué, which is simply the glueing or sewing of pieces of cloth on a background to create a design. Casein-base fabric adhesives are excellent for fabric pictures, but remember, if your picture must be washed often this glue is not permanent. You can also use white glue. Thin plastic materials are available which you can iron in between two pieces of cloth to hold them together, eliminating the need for sewing. If you do choose to sew, there are temporary glues available for spraying on the cloth pieces to hold them flat while you sew the edges of the cloth to the background material. Since these glues are not permanent, you can easily pull the fabric off the background and place it in another position if you need to.

Illus. 137 is a stunning collage incorporating scrap fabrics with other materials on a canvas-covered board. The colorful circus tents and flags are made of pieces of red and white felt which are glued to the canvas. Strands of yarn form the flagpoles and the trim on the tents. The seal and the ball and box that he is balancing were cut out of pieces of scrap sheet cork and glued to the canvas background. Very small nails, pushed into the cork, formed the frames for the colored-string designs. (See page 76.)

Illus. 138 (left) and 139 (right). The people collages on these pages were made by elementary-school children. Here a satin remnant provides an elegant gown and a pair of gloves for the lady and odds and ends of fabric patches make up the outfit for the tattered tramp.

Illus. 140. These two fabric-collage portraits were created by fifth-grade students. The lady has a pheasant-feather hat and scouring-pad hair. The elderly fellow on the right has a cotton batting (cotton wool) beard, and yarn eyebrows. Lace scraps decorate his cloth hat.

Illus. 141. Heavy textured white and black fabrics provide the formal outfit for this heavily bearded figure.

Illus. 142. A scrap of a furry cloth material is ideal for this lady's coat. Imitation fur serves as trim for the bottom of her skirt.

Illus. 143. Whoever receives one of these fabric collage greeting cards is certain to feel special. Scrap pieces of colored burlap (hessian) are glued onto white paper as backgrounds for all three cards. Small pieces of felt, cut into flower and leaf shapes, are glued to the burlap in the card at the left. Felt leaves and bead berries form the wreath on the Christmas card in the middle. A large poinsettia with felt petals, metallic cord trim, Christmas glitter, and small yarn decorations make up the cheerful card at the right.

The doll house in Illus. 144 shows the fabric collage technique applied to a three-dimensional surface. A variety of double-knit fabric scraps were glued onto a corrugated cardboard box which had been cut down to the house shape. In Illus. 145, you can see that the top of a box was cut off, leaving four sides and a bottom. Then the sides were cut as shown to accommodate the roof, which is simply made by bending over a large sheet of corrugated cardboard. The chimney is a piece of cardboard bent into a cube shape and inserted into a square hole cut into the roof.

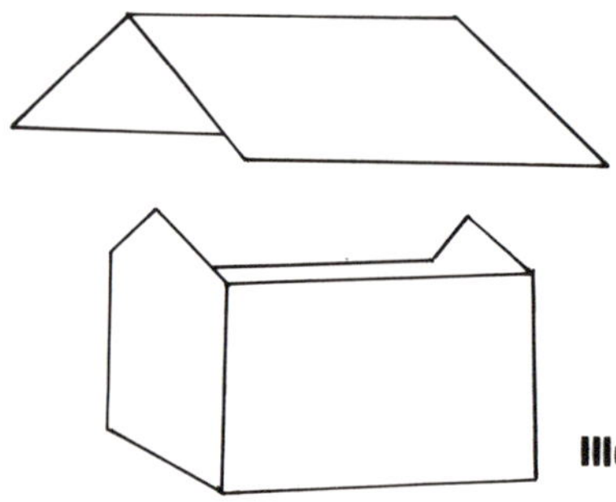

Illus. 145.

Illus. 144 (left). This charming appliquéd doll house was constructed from cardboard as shown in the drawing above.

Illus. 146. In "Harvest Time" brown burlap was stapled onto a white pine frame of 2 × 2-inch (5 × 5-cm.) wood. Felt pieces were all sewn onto the burlap with rug yarn, 4-ply yarns, and black thread.

Illus. 146 and Illus. 148 are two examples of sewn appliqué fabric collages. When sewing fabrics onto fabric backgrounds, the blanket stitch (Illus. 147) is ideal. It is easy to do and offers an attractive edging. First spray temporary glue (see page 55) on the piece of fabric so that it lies flat when put on the cloth background. Illus. 147 hows how the needle and thread come up through the background material (#1) and go back down into and through both the background material and the appliquéd piece, forming the letter "J". When coming back up through the background material the second time (#1), pull the needle through the curve of the "J" and down through the appliquéd piece (#2).

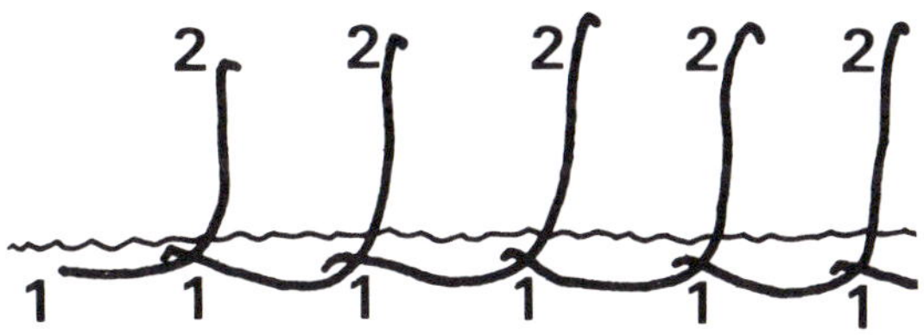

Illus. 147. The blanket stitch.

Illus. 148. For this orange tree felt pieces form the leaves, and fruit and yarn, the trunk and branches.

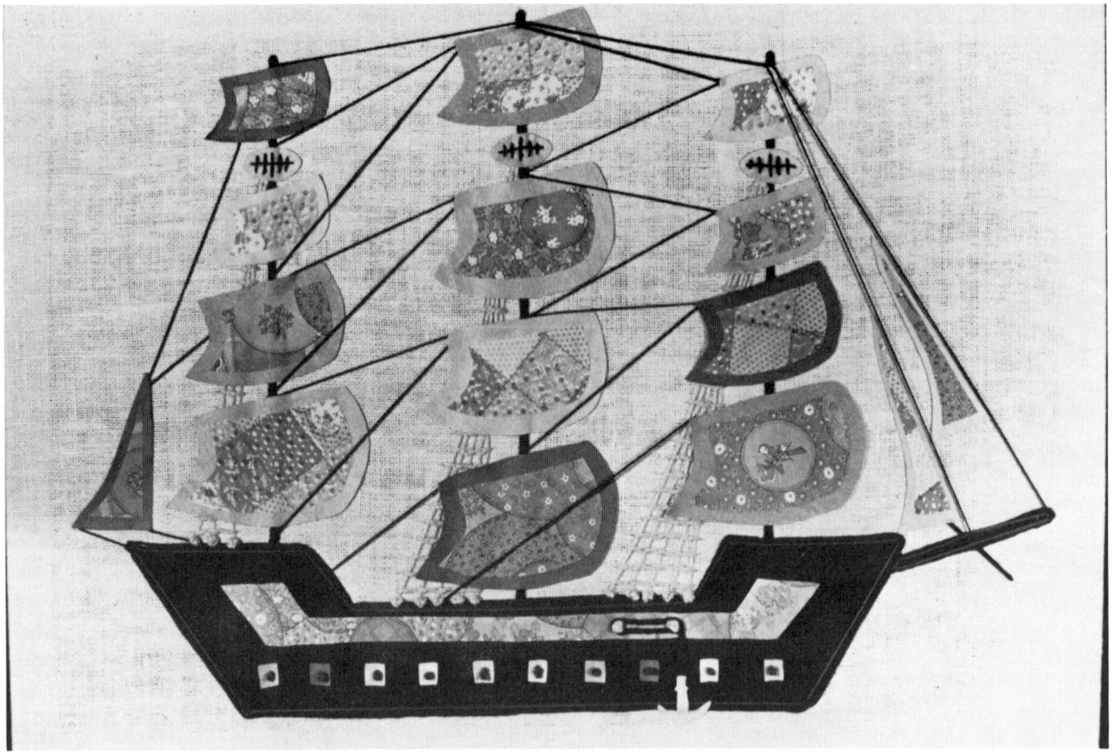

Illus. 149. The attractive sails on this old ship are composed of many different patterned scraps of fabric appliquéd on pieces of felt with both the blanket stitch and glue. These are in turn appliquéd on a burlap background with glue. The hull of the ship is made of a large piece of felt, which has small windows and trim of appliquéd fabric. Strands of yarn form the "lines" of the sails.

Illus. 150. In this striking design, pieces of Moroccan wool are appliquéd onto a large piece of wool with machine stitches.

Illus. 151 (above) and Illus. 152 (left). These geometric designs are composed of plain-colored Mexican cotton fabric remnants machine-stitched together and applied to a background board with glue. (Courtesy of Margaret Yuill.)

Illus. 153.

Illus. 154.

These three fabric collages are all composed of machine-stitched pieces of fabric, but a unique method was used to create the effect you see here. The sewing machine was set for the straight stitch and run back and forth over the fabric collage continuously until a thick layer of threads was built up creating an almost impressionistic appearance. (Courtesy of Margaret Yuill.)

Illus. 155.

Illus. 156. Scrap fabrics and yarns were used to create this striking group on a burlap background. The figure on the left is wearing a leather neck piece, while a piece of twine forms the belt for the friend next to him.

The fabric collage in Illus. 156 is a combination of stitching and sewn and glued appliqué. All four figures have hair created from scrap yarn which was pulled apart and sewn and glued to the off-white burlap background. A piece of angora yarn provides the fur coat for the poodle. Various scrap fabrics—corduroy, denim and leather—were used for clothing. The denim overalls on the figure at the far right sport machine-stitched line designs. The fellow on the left is wearing a chain-stitched sweater. Illus. 157 shows how the chain stitch is made. The yarn is brought up through the burlap (#1) and back

Illus. 157. The chain stitch.

down again (#2) to form a loop. Each successive loop comes up and back down again *within* the last loop made.

The pastoral scene in Illus. 158 is composed completely of stitchery, in this case the satin stitch. A #13 tapestry needle was used to pull the yarn remnants through the burlap background from back to front and to the back again, creating the long stitches making up this beautiful scene.

Illus. 158. Although the satin stitch is used exclusively in this rural scene, the green meadow in the middle shows the interesting effect of combining horizontal and vertical satin stitches. This is an ideal technique for using up small pieces of yarn because the stitches can be tied up easily on the reverse of the backing material.

Illus. 159. "The Dance of the Warriors" is made up completely of the satin stitch and utilizes the burlap background itself as part of the design. Scraps of different-colored yarn were used throughout.

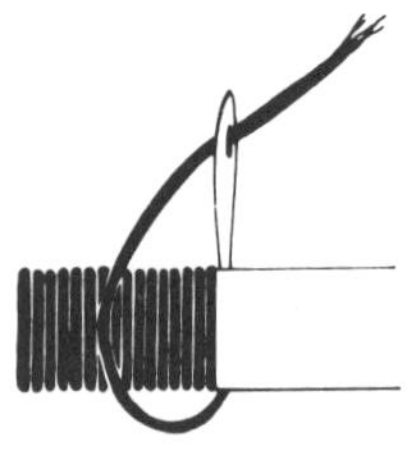

Illus. 160. The satin stitch.

Illus. 161. A piece of brown burlap is appliquéd on the white burlap background to provide the water wheel for this stitched rustic scene. Real stones form the foundation of the structure and are outlined with scrap yarn for emphasis.

Illus. 162. Off-white burlap provides the backing here. The geometrical outlines of the design are composed of black rug filler. Most of the shapes are made up of simple straight stitches, while others are appliquéd felt. The three light-colored circles in the middle were created with the couching stitch (long loops made by simply stitching in and out of the burlap loosely). The ends of the loops were clipped with scissors.

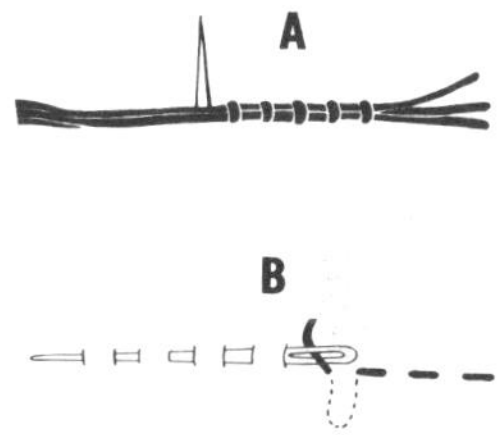

Illus. 163. A. The couching stitch. B. The straight stitch.

Illus. 164. The cross stitch.

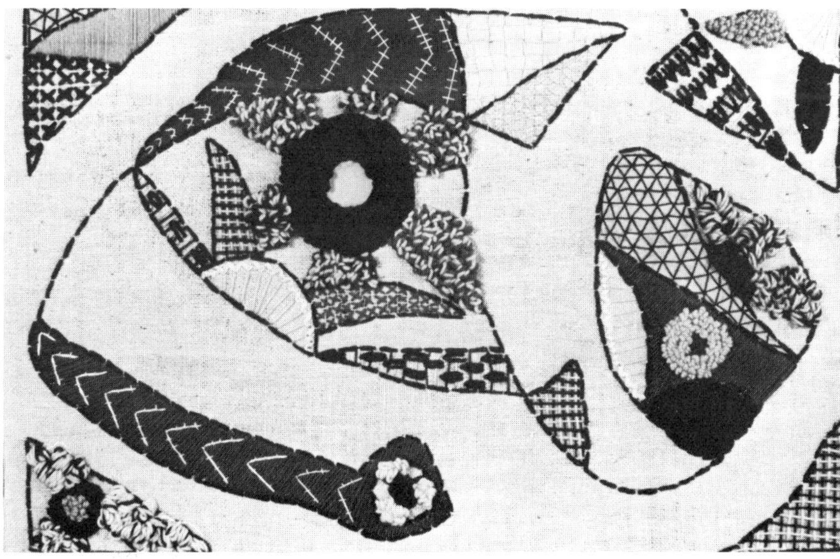

Illus. 165. A simple scribble drawing is the basis for this stitchery design. Pieces of felt and rick-rack are appliquéd onto off-white burlap. The couching stitch (see Illus. 163), satin stitch, and cross stitch were all used in various areas of the design.

Illus. 166. This unusual landscape shows how simply you can create trees with long and short stitches.

Illus. 167. The long and short stitches.

The unique landscape in Illus. 166 has a sky created by sewing heavy yarn over pencils and wire which were later removed. The dark hills in the background are composed of French knots and the trees in the middleground of vertical long and short stitches. In the lower half of the landscape, the weft threads of the burlap backing were cut and removed and beach grass was woven loosely in their place into the warp threads. Small ornamental glass circles and pieces of felt complete the picture.

The exciting sunrise in Illus. 168 is happening behind a grey mountain made of a piece of appliquéd fabric. French knots make up the four triangular corners and the satin stitch is used throughout the rest of the scene—in the sky, the sun, the land and the trees.

An exciting and unusual theme for a stitchery—an old house—is shown in three different renderings in Illus. 169, 170, and 171. The intricate details of Victorian houses provide a multitude of possibilities for all kinds of small pieces of scrap fabrics and yarns, as well as other kinds of materials. For example, roofs might be constructed of pieces of metal or aluminum or copper foil, shutters and doors of small pieces of wood, fences of wire, and so on. Even artificial flowers could be incorporated into the lawn surrounding the house. A bit of

Illus. 168. "Sunrise at San Gabriel."

Illus. 169 (left). Felt provides the basic element for this charming old house. Dark brown felt creates the roof and stone path and red felt the chimney and draperies. The entire façade of the house is stitched with yarn. The shrubbery's rough texture is furnished by French knots. Illus. 170 (right). The façade of this house is made up of the satin stitch. Pieces of ribbon and felt compose the draperies and curtains as well as the "etched-glass" window in the front door. The heavy black iron fence was made by sewing yarn over wire, which was then left inside the yarn for support. A tooled aluminum roof (see page 105) and felt clouds put the finishing touch on the scene.

sandpaper might serve as a chimney and plastic wrap could be used for the windows. Once you get started on a stitchery house, you will find the possibilities are unlimited!

Illus. 171. "The Hackley House," located in Grand Haven, Michigan, served as the inspiration for this fine fabric picture. Tooled scrap copper adds a bright touch across the middle and around the porte-cochere, while leather remnants decorate some of the roof sections. Appliquéd felt and the satin stitch make up the remainder of the house.

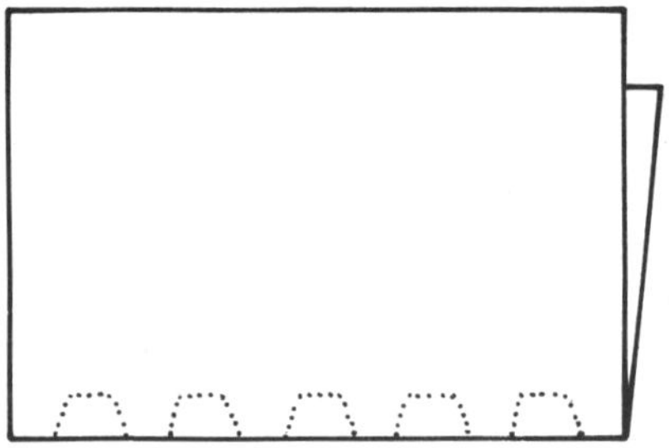

Illus. 173.

Illus. 172. This unusual design is a two-layer window stitchery. The upper layer of fabric is cut into various shapes along folds as shown in the diagram above.

Another technique you might want to try is a two-layer "window" stitchery, which is very easy to do and allows you to experiment with interesting accidental designs. You will need two different-colored pieces of fabric the same size. Illus. 173 shows how the top layer of fabric is folded and "windows" cut out along the fold. The shape shown here would create a six-sided shape, or a hexagon. A "V" would result in a diamond shape, a half-circle in a full circle. Experiment with different cuts on a piece of folded paper if you want. To make rows such as shown in Illus. 172, simply keep folding the fabric at intervals and cutting. (You might also want to mix cuts along one fold.)

In Illus. 172, a variety of shapes were cut from black burlap. A piece of chartreuse burlap was stapled to a wooden frame and then the black design stapled on top of it. Here, small beads are sewn between some of the windows.

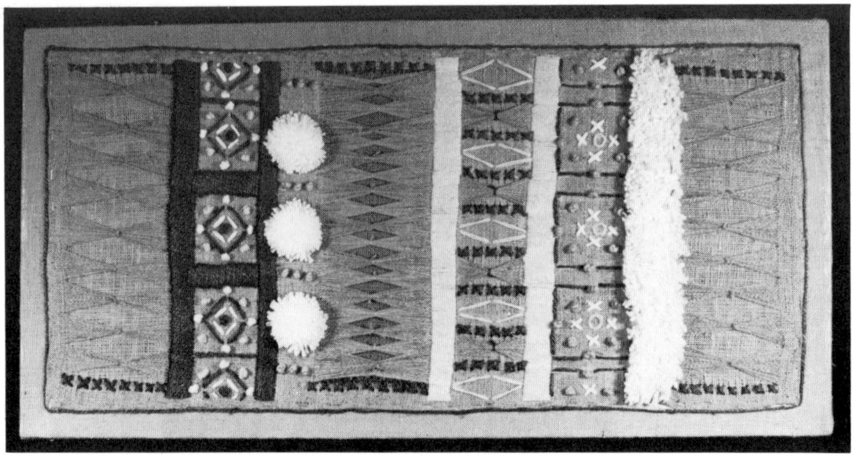

Illus. 174. Another two-layer window stitchery displays "unweaving," that is, pulling the weft threads out in certain areas rather than cutting out patterns.

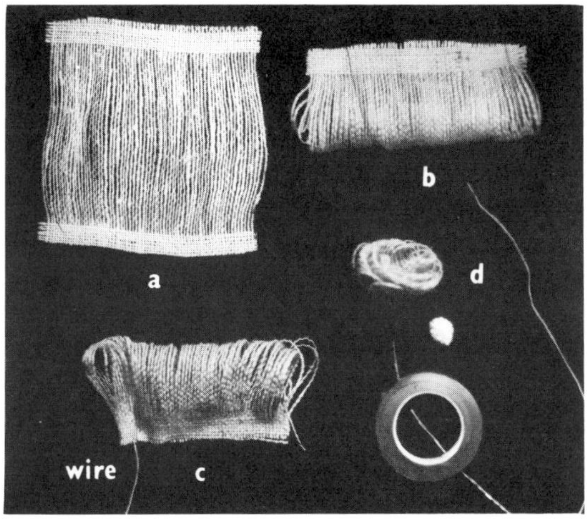

Illus. 175 (left) and 176 (right). These are the steps in creating the two different kinds of colored burlap flowers shown in Illus. 177.

If you use burlap, instead of cutting out the shapes from the top layer, you might remove some of the weft threads as shown in Illus. 174, and then tie the remaining warp threads, creating diamond-shaped windows. In this example, green burlap was layered over orange. A variety of different stitches in red and gold yarns ornament the basic window design.

All you need is some pieces of scrap burlap to make a beautiful bouquet of flowers like the one in Illus. 177! To make the flowers, first cut out a 7-inch (17.7-cm.) square from colored burlap. Then pull out all of the weft threads so that only seven remain on both the top and the bottom as shown at *a* in Illus. 175. Fold the square so that the top and bottom edges line up (*b*). Place a piece of wire on one end of the folded-over burlap (*c*) and roll up. Wrap tape around the base where the threads were not removed. Then plump up the blossom with your fingers (*d*). Make as many as you want in different colors.

Now take the removed threads and make a blossom with them, too. Bunch the threads together and tie wire around the middle as shown on the left in Illus. 176. Press the two sections tightly together and wrap florist's tape around the base and the wire as shown at the right.

Illus. 177. Orange, brown, and yellow burlap scraps were used to make this charming floral arrangement. See the text and the two photographs above for directions on how to make it. See another arrangement in color on the next page.

Illus. 178. To make a colorful burlap floral bouquet such as this, see page 69.

The two charming puppets shown in Illus. 179 and Illus. 180 are a wonderful way to use up all kinds of fabric scraps, plus a variety of other materials.

"Lavern" in Illus. 179 has a cylindrical saltbox foundation, around which a large piece of burlap has been glued. The top and bottom of the box have been removed. Red yarn was sewn onto the burlap to form the hair and the fringe at the bottom, and to suspend the wooden bead earrings. Plastic foam ears and eyes were cut out of an egg carton. A bead necklace and old pair of play eyeglasses complete this hand puppet.

An old knitted sweater sleeve serves as the soft body for the hand puppet in Illus. 180. To firm up the head, two plastic drinking cups were stuffed up the sleeve. An old shoelace defines the head area. Blue-felt eyes, a red-felt mouth and a brown-felt hat add the finishing touches, and your puppet is ready to perform!

Illus. 179.

Illus. 180.

Illus. 181. Christmas glitter is being applied to the cover of this colorful woven container.

Illus. 182.

Scrap weaving presents tremendous possibilities, because you cannot only use scrap materials to weave with, but you can use scrap looms—cartons, branches, drinking straws, to name a few. An example of carton-loom weaving is shown in Illus. 181. Here, a circular carton and its cover are being given their finishing touches.

Illus. 182 shows the beginning of the weaving of the cover of a circular carton. Evenly spaced slits are cut into the rim of the cover which has had a piece of grey construction paper glued onto it. If the weaving is loose, this will show through. Next, a piece of yarn is inserted into the bottom notch, or notch #1, and pulled across the top of the cover into the notch directly opposite, notch #2 (Illus. 183). Then the yarn is moved into the notch on the left of #2, notch #3, pulled across the bottom and threaded into notch #4, which is immediately to the right of #1 (Illus. 183). The process continues, always going in a counter-clockwise direction.

Illus. 183.

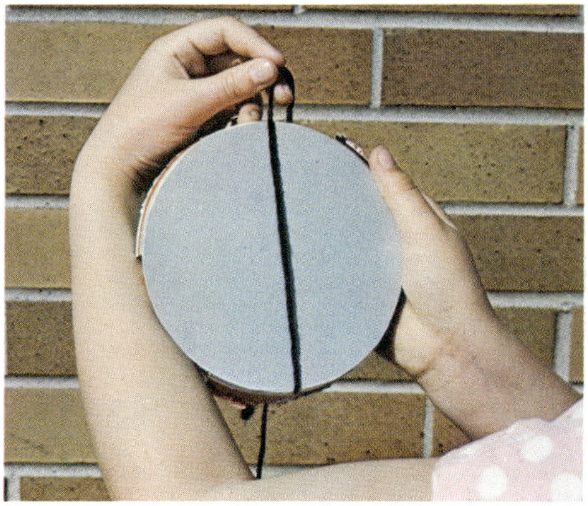

Illus. 184.

Illus. 185. Thread the weft thread on the outside edge of the cover and pull the threads to the middle.

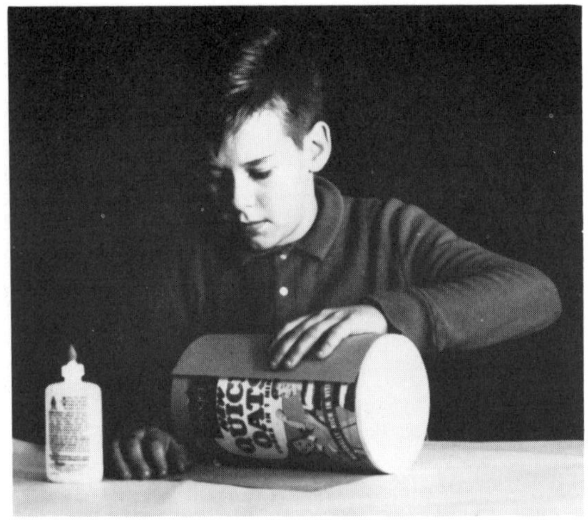

Illus. 186. Cover the round container with colored construction paper before weaving.

Once the threading of the warp is finished, you are ready to weave the weft. It is easier if you insert the weft thread along the outer edge of the circle and then push or pull it in towards the middle (Illus. 185). Continue this until you have completely finished the cover.

After having woven over the circular cover, you should have no difficulty weaving on the round container. First, glue colored construction paper onto the container so that in case it shows through the weaving, it will look like part of the design. Cut evenly spaced notches not more than $\frac{1}{2}$ inch (1.2 cm.) deep into the rim for threading the warp.

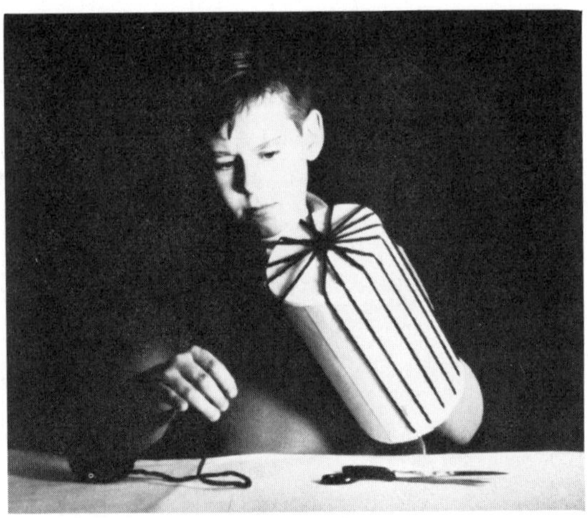

Illus. 187. Make sure that the warp threads cross each other at the middle of the bottom.

Illus. 188. Begin the weft threading at the bottom of the container.

Illus. 189. Two milk cartons were used for looms for these woven creations. Yarn flowers are glued to the sides for decoration. Illus. 190 shows the weaving pattern.

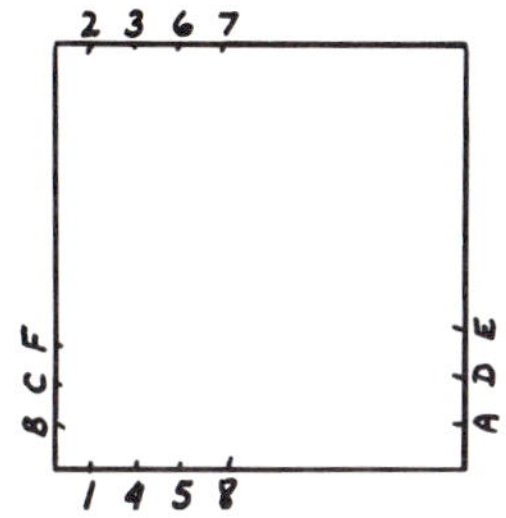

Illus. 190.

Insert the thread into the notch directly in front of you (notch #1), bring it down the side, across the bottom, and up the other side into the notch directly opposite (notch #2). Proceed counter-clockwise just as you did with the cover until the notches are all filled (Illus. 187). Then start weaving the weft thread in and out, beginning at the bottom and weaving up (Illus. 188).

When using a four-sided loom such as the milk carton in Illus. 189, follow the weaving pattern shown in Illus. 190.

An example of radial weaving is shown in Illus. 192. First draw a circle on heavy cardboard or plywood which is used here. Punch or drill evenly spaced holes round the edge of the circle. Starting in any hole, run the yarn or string to a hole directly opposite. Insert and bring up through the next hole going in a counter-clockwise direction (Illus. 191).

Illus. 191. This is the weaving pattern for the radial design in Illus. 192.

Illus. 192. A variety of weft materials were used here—yarn, twine, feathers and twigs.

Illus. 193. An old lampshade frame here gets a new look. Warp threads were wound round the top and bottom rims. The weft is formed of horizontal rows of colored scrap yarns, woven loosely to produce an open, airy design. The fringe at the bottom consists of strands of yarn knotted onto the rim.

Illus. 194. "Sherman's Throne #1" has a loom composed of an old toilet seat. Staples on the reverse side of the seat support the warp threads which were run to a circular metal washer in the left middle. Beginning round the washer, the weft threads were woven in and out of the warp in a radial pattern.

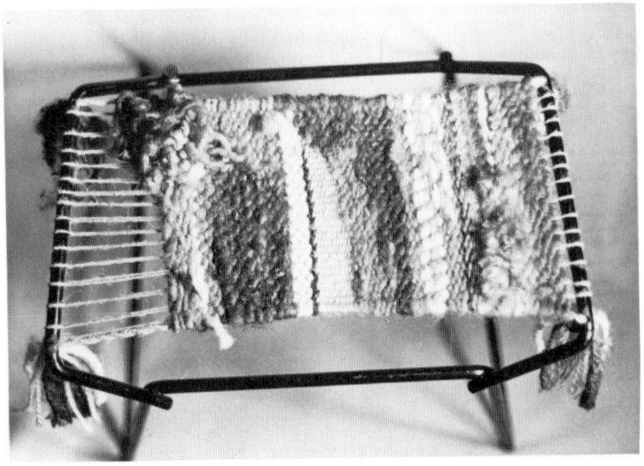

Illus. 195. "Montezuma's Bench" makes use of another unique loom—a luggage rack that had a decrepit covering. Here the covering has been removed and a linen yarn warp wound round opposite ends of the frame. The weft consists of spun silk, wool yarn and acrylic yarn.

Illus. 196. This unfinished weaving is being created on an old grill used for cooking on an open fire. In the middle of the design, the yarn is woven in and out of the wires in a radial fashion (see page 73 on radial weaving).

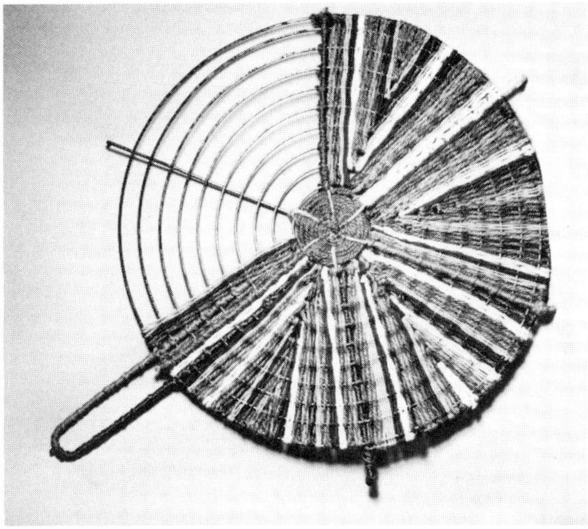

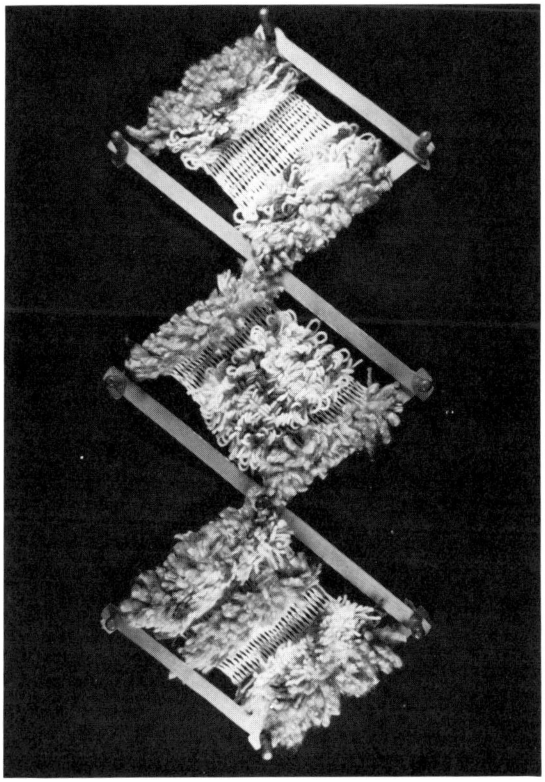

Illus. 197. A wooden clothing rack provides the loom for this weaving. Nails were pounded into opposite wooden sections to thread the warp. The weft consists of Mexican wool.

Illus. 198. If you happen to have some deer antlers hanging round the house, look what you can do with them! Just nail them to a straight tree branch or piece of driftwood which serves as the loom.

Illus. 199. The spider and his web are admirably set off against a glass "window" backed by black construction paper.

Illus. 200. There is no one method for drawing with nails and yarn. You will want to experiment with different patterns and methods.

It's hard to believe the fabulous spider picture in Illus. 199 is made from little more than string and yarn. A rectangular window was cut out of yellow cardboard to create the mat. Holes were punched into the mat at intervals and pieces of elastic string were inserted into the holes, run to the middle and tied round a metal ring, which forms the foundation for the spider's body. More elastic string was used to create the webbing, each piece being tied to the radiating strings. A pipe cleaner, shaped into a circle forms the body and covers the ring. A small ball of yarn was glued inside the pipe cleaner. Pipe cleaners poked into the body form the legs.

A different kind of "drawing" is shown in Illus. 200. Here only yarn and nails are needed. Simply drive nails into a wooden backing (nails with small heads work best) after drawing a pattern on the board with pencil. Tie the yarn on a nail and then make your drawing by winding it round the other nails. When you want to change color, simply tie off the yarn on a nail and continue with the new yarn.

Illus. 201. A three-dimensional string and yarn drawing is easy to do inside a painted corrugated cardboard box. Simply lace the string through holes in the sides. Beads were added here for a decorative touch.

The handy scrap box in Illus. 202 has a metal lard can foundation. A piece of scrap carpeting is wrapped round the can and attached with carpet tape. Jute (the fabric in burlap or hessian) was used to form the diamond design over the carpeting. Simple square knots tie the two colors together.

Two yarn mobiles are shown in Illus. 203 and Illus. 204. To make a two-dimensional one, soak your yarn in wheat paste and lay out your design with the yarn on wax paper. Allow to dry and then remove the paper. The yarn should be stiff enough to hold its shape. Cut pieces of wire coat hangers to suspend the mobile and then attach the design to the wire pieces with string and hang your mobile from the ceiling.

The three-dimensional mobile in Illus. 204 is made slightly differently. The yarn is soaked in wheat paste as before. Then, forming your design as you go, place the yarn over balloons and allow to dry. When thoroughly dry, prick the balloons with a pin and remove them.

Illus. 202. Jute, a carpet remnant, and a crocheted doily here transform a metal lard can into a what-not box.

Illus. 203.

Illus. 204.

Wood

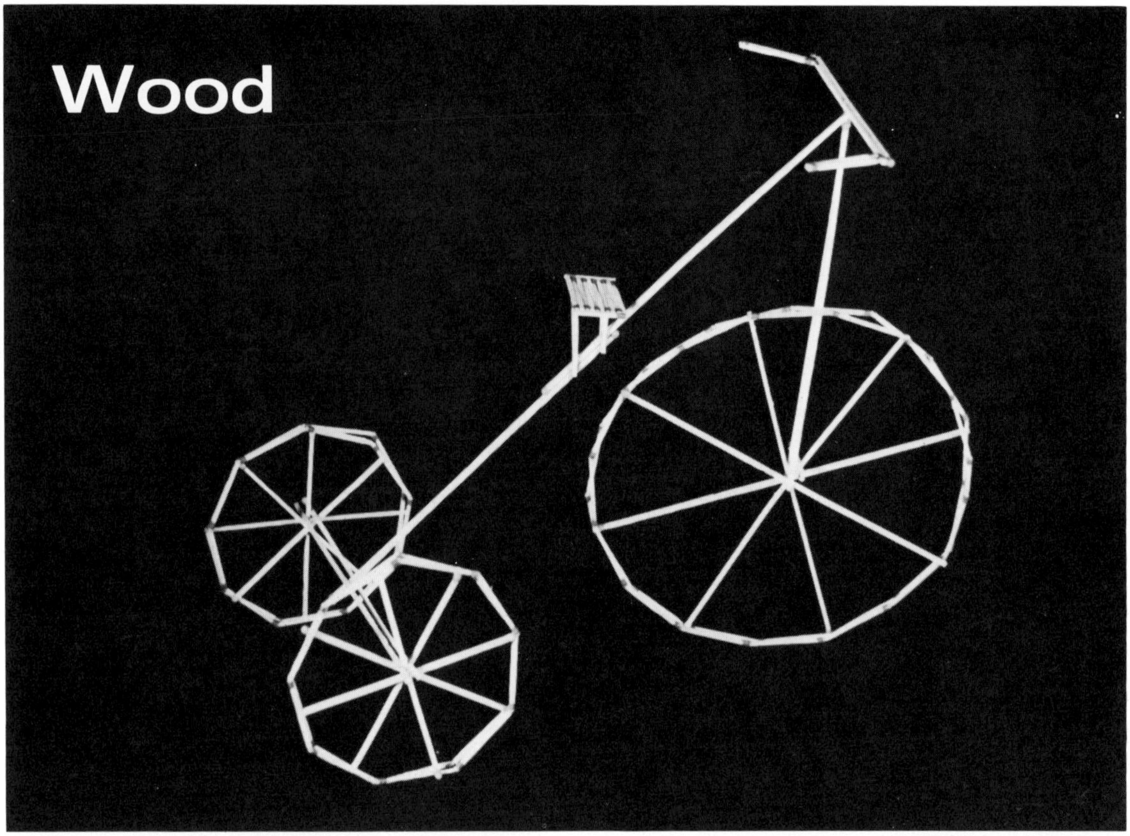

Illus. 205. This charming old-fashioned tricycle is made with wooden matchsticks.

Wood can be found everywhere. Just look round you right now and see how many things you can find that are made of wood. Surprised? No matter where you are—indoors or out-of-doors, you will find some form of wood, from matchsticks to buildings, twigs to logs. Drift-wood, wood shavings, branches, boxes, crates, bark, old broken furniture, toothpicks, ice-cream sticks, spools, carpenters' scraps, wooden buttons, broken toys, bamboo, pencils, children's building blocks, broom handles, cigar boxes, wooden doorknobs, bed slats, old picture frames—these are just some of the multitude of sources for your wood-scrap crafting.

You will find that most wood is so attractive in its natural state, that you won't want to hide it under paints and other decorations. However, you might want to emphasize its grain by running an acetylene torch over the surface, or using wood stains, shoe polish, or boiled linseed oil. Furniture or floor wax provide a protective finish and at the same time preserve the natural beauty of the wood.

Although many of the project suggestions in this chapter require little more than glue, you will find a hammer and nails and a screwdriver and screws handy for some of the other projects. If you have the equipment handy, you can try some of the more ambitious projects that call for handsaws or electric saws such as the jigsaw or the coping saw. Soft woods such as balsa, pine and basswood are easily sawed with hand tools, while hard woods—oak and maple, for example—require electric saws.

In many cases, you will find carving knives and chisels useful for adding texture and patterns, as well as sandpaper for smoothing purposes. Ordinary white glue will do for most of your glued projects; however, if your object is to be used or displayed out-of-doors, you should use epoxy cement or waterproof glue.

The wooden matchstick creations shown here are easy to do, require only glue and a supply of *burnt* matches, so why not get started on your wood crafting with a matchstick project!

The antique tricycle in Illus. 205 is constructed of regular-sized matchsticks and the extra long kind used for lighting candles in deep containers. The first step is to glue (with quick-drying plane glue) the wheels together. Then glue the cross-bar and the handlebars to each other. Let all the parts dry by leaning them against a stack of books. Then assemble the body and the wheels and let dry again.

Illus. 206 shows a match-covered vase. A quart-size milk carton was cut down to serve as the foundation, and the matchsticks were glued on in an interesting design with white glue.

Illus. 206.

Illus. 207. Here a matchstick collage was created by glueing matchsticks onto a background of driftwood.

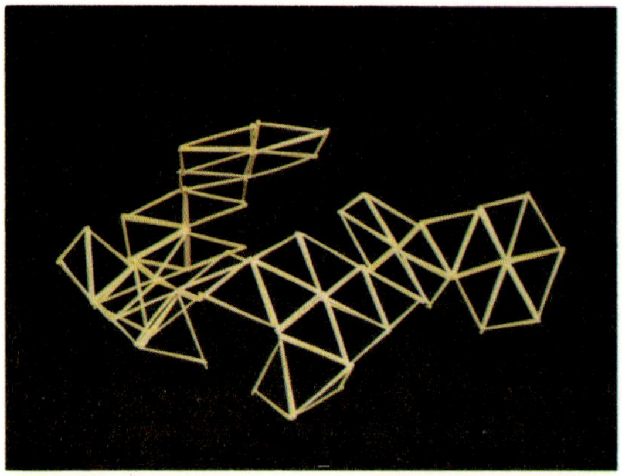

Illus. 208. Instead of using matchsticks, you can make a toothpick sculpture like this. Here a free-form sculpture was made from glued triangles.

Illus. 209 (right). Here, flat wooden toothpicks make a more elaborate sculpture which is glued to a scrap-wood base covered with felt. Aluminum spray paint was applied to the sculpture before glueing it to the base, lending it a weighted structural feeling. Bridges, buildings, as well as free-form sculptures can all be created with wooden toothpicks.

Odd pieces of broken furniture provide ample material for creating all kinds of fantastic objects as you can see in the table-leg sculpture in Illus. 210. Chair legs, newel posts, bedposts, and finials on all kinds of furniture have delightful shapes to decorate, without having to be cut or sawed. Small pieces of rope here serve as arms for all three figures. The figure on the left has wood-shavings hair and black elastic-thread eyebrows. The middle figure has a binder-twine beard and the small figure has curled copper-wire hair and a black elastic-thread nose. Sequins serve as eyes and lace as collars for the two outside figures.

Illus. 210. "The Aristocrats" are composed of table legs and pieces of stair balusters. Their decorations consist of a number of small scrap objects attached with white glue.

Illus. 211. Chair legs and stair balusters provide bodies for dignitaries whose faces were cut from magazines and applied to the wood with Decopage.

Illus. 212. A nursery-school child constructed this vehicle from wood scraps, all of which are held together with white glue.

Illus. 213. A torch was run over the plywood base of this sculpture to accentuate the wood grain. Scrap furniture pieces were glued on the base in an abstract design.

Illus. 214. Pieces of Masonite (pressed-board) provide the foundation here for a collage. Scrap wood is nailed on in various areas and scrap metal pieces are adhered with epoxy cement. Added color and decoration is made by acrylic paint.

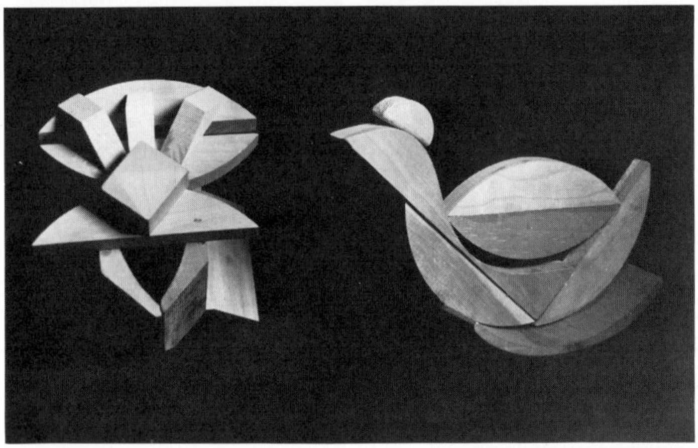

Illus. 215. Scrap wood from a furniture factory is glued together to create two stunning birds—a turkey on the left and a duck on the right.

Illus. 216. In this wood bas-relief, left-over carpenter's wood is glued to a large piece of white pine stained with a fruitwood stain. The blocks consist of pine and redwood. The redwood was stained with maple stain, and the pine with walnut stain. Some of the pine pieces were left in their natural state.

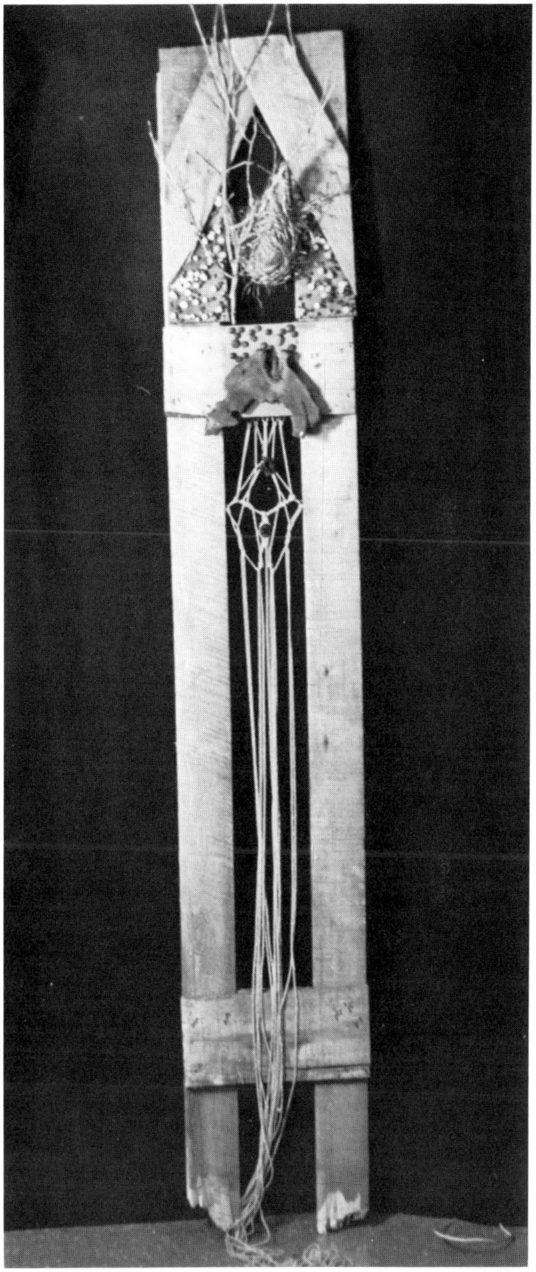

Illus. 217 (left). This 6-foot (2-m.) high sculpture is constructed from scrap mahogany pieces obtained from a wood-working company. Nails and glue hold the pieces together. Illus. 218 (right). Looking very much like a grandfather clock, this 7-foot (2.3-m.) sculpture has a basic frame of old wood boards nailed together. An old bird's nest and branches add decorative touches. White string dyed in water containing tea bags, hangs between the two vertical boards.

Illus. 219. "Back Home Again in Indiana" is a shallow diorama created inside a wooden seedling box.

The diorama in Illus. 219 was made in a wooden seedling box. Plastic foam, covered with moss, serves as a floor and burlap the back wall. Dried flowers create an effective garden for the corn-husk figures who are busy at work. Small pieces of wood were used to create the door, the fence and the birdhouse, all of which were either stained or painted. Scrap bits of ribbon surround the window on the door to give the impression of curtains. A tiny toy milk container holds a big bouquet of straw flowers at the far right. White glue was used to hold all the materials together and to the background.

Illus. 220. Duck and pheasant feathers glued to scrap wood pieces create a fine owl with glass marble eyes.

Illus. 221. A large wooden electrical-wire spool discarded by a power company makes an ideal end table or stool. The wood was sanded, covered with white latex paint, and then painted with bright acrylic paints.

Illus. 222. Scrap wood constitutes the material for this A-frame house which measures 2 feet wide × 18 inches high (60 cm. × 45 cm.). Rubber cement (Cow), epoxy cement, and nails serve as connecting materials for the bark and other wood pieces, as well as the pine cones and pine needles.

The charming street scene in Illus. 223 can be adapted in any way you like, depending upon the sizes and shapes of the wood you have on hand. Here, a large piece of scrap wood serves as the base upon which wood-block buildings are glued. First the wood blocks were painted to look like houses and glued onto the base. Next, a long wood strip was glued on as a sidewalk. The lamp-posts, consisting of long nails with wooden beads glued on top, were pounded into the base. Small branches are glued on as trees.

Illus. 223. This street scene was created with scrap wood blocks and a number of decorative materials. The "grass" around the lamp consists of hemp and "road" of aquarium gravel. The shutters and roof on the left are felt, and the large central roof is sandpaper.

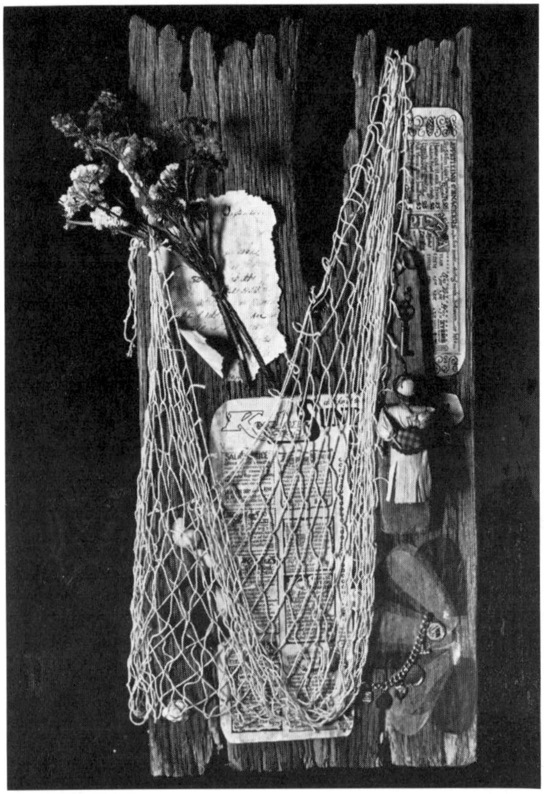

Illus. 224 (left). "Memory Lane" is the title of this collage on old barn wood. Pieces of driftwood, a restaurant menu, a corn-husk doll, a key, shells, leather, fishnet, dried flowers, an old letter (painted with coffee for an antique effect) are all glued in place. Illus. 225 (right). "Treasure Board" has a wooden backing and lathe-turned porch balusters forming a frame for sea shells, a hygrometer, and old knobs.

The dresser drawer in Illus. 226 was divided up by wood boards ½ inch thick by 4½ inches deep (1.2 cm. by 11.4 cm.), cut to the desired widths with a handsaw. These boards were stained with a fruitwood oil stain, as was the drawer itself, and nailed in place.

The sections of the ecology box (Illus. 227) were filled with spiral macaroni, a wooden trivet, gourds, oats, field corn and split peas. U-shaped

Illus. 226. An old dresser drawer provides an ideal background for an ecology box.

Illus. 227. Here is the dresser drawer shown in Illus. 226 transformed into an ecology box, all filled with seeds and such and ready to be mounted on the wall.

pieces of corrugated cardboard were inserted into some of the compartments and then the grains were glued to them. This created a hollow area behind the seeds and lightened the weight of the box, at the same time creating the illusion that each compartment was filled completely with seeds. A piece of sheet plastic was cut to fit over the front of the box and then the box was fitted with strong picture hooks and hung on the wall.

Illus. 228. A free-form sculpture is here created with old pieces of wood. Some of the pieces are nailed together and others are attached with wood dowels inserted into drilled holes.

Illus. 229. "Time" is the title of this collage. White glue and epoxy cement hold a multitude of watch and clock parts to the barn wood base.

Illus. 230. Wooden thread spools are fun to experiment with. Here are Dorothy and the Tin Man from "The Wizard of Oz." Wire inserted through the spool holes is turned and twisted to provide bodies, legs, and arms. All kinds of small scrap material can be used to make clothing and other adornments.

Illus. 231. Plywood is here painted with black India ink to form the background for an eggshell mosaic. The design was first formed with yarn which was glued to the base, and then broken eggshells, dyed with food coloring, were glued within the resulting spaces.

The handy sewing organizer in Illus. 232 is made from a scrap wooden board that was sanded, stained, and varnished. The tops of baby-food jars were painted and nailed, top-side-down, to the board and the jars themselves were filled with sewing needs and screwed onto the tops. A ruler for measuring is nailed to the upper edge, and a nail holds a hem gauge on the left. The scissors is housed in two picture hooks, while a tape measure hangs from another hook on the right.

Illus. 232. A wooden board is used to make this handy sewing organizer.

Illus. 233. This jolly little train engine and coal car are made of pieces of scrap wood, some of which were sawed to shape. Discs, sawed from dowelling, provide the wheels. A screw eye and eyelet make the coupling. The little engineer is made from wood scraps.

"Movement in Growth" is a window flower box decorated to look like a car. Constructed of $\frac{3}{4}$-inch (2-cm.)-thick plywood the long sides measure 30 inches wide by 6 inches (76 cm. by 15 cm.) high. Two aluminum soda cans glued to the bottom serve as wheels, and red checkers (draughts) are nailed in place for tail-lights. The headlights are Christmas lights. A faucet handle mounted on a pencil provides a steering wheel and a piece of aluminum tubing an exhaust pipe. A pulley belt from a car was cut up into four pieces and glued on to make fenders. Then the car was decorated with paint and nylon cord to suggest flowers.

Illus. 234. This window box is decorated to look like a car with soda-can wheels and a faucet steering wheel.

The Tiki god in Illus. 235 was made from a piece of knot-free white pine, measuring 4 × 4 × 24 inches (10 × 10 × 60 cm.). This project involves a table saw, a dado head (a

Illus. 236. A 9-year-old child made this African mask from old furniture wood scraps, all of which are glued together with epoxy cement. The basic shape was cut from a piece of wood with a coping saw. Wood shavings are glued on to make a moustache and pieces of wood dowelling serve as earrings. The hair is composed of a large piece of rope bound with electrical tape.

chisel can be used instead of the dado head), a hand saw, a coping saw (for cutting eyes and nose), and a wood chisel (for taking off rough edges). Cuts were made into the wood with the table saw and dado head. Angles were cut with

Illus. 235. This elegant Tiki god is constructed from one piece of wood.

Illus. 237. These are the various parts of barn wood all cut and ready to be assembled into the wall planter in Illus. 238.

Illus. 238. You will need a jigsaw and a hand drill and screws to make this planter. Since barn wood is very brittle, it is wise not to use nails.

the hand saw. The eyes and nose, cut with a coping saw from scrap wood, were glued in place. Thumbtacks serve as the pupils of his eyes. Finally, the entire statue was stained.

The charming wall planter in Illus. 238 was constructed out of old barn wood and sawed with a jigsaw. The parts that you will need are shown in Illus. 237. The back panel is 26 inches high by 10 inches wide (66 cm. by 25 cm.). The circle, removed with the saw, is 7 inches (17 cm.) in circumference and measures $2\frac{3}{8}$ inches (6 cm.) in from each side and $7\frac{1}{2}$ inches (18.7 cm.) from the top. The bottom piece measures 10 inches (25 cm.) along the back and $10\frac{1}{2}$ inches (26 cm.) in the front, tapering slightly from back to front. The sides of this bottom piece measure $9\frac{1}{4}$ inches (23 cm.). The front piece is 4 by 8 inches (10 by 20 cm.).

Since the weathered barn wood is very brittle and could be split by nails, holes were drilled in it, and counter-sunk screws were inserted to hold the pieces together. Here and there some white glue was also used for added reinforcement.

Illus. 239. All you need are a base of some kind, small pieces of wood flooring, and a piece of plywood to make this table.

An old wooden folding stool became the base for the table in Illus. 239, which you can make easily. A piece of $\frac{3}{4}$-inch (1.87-cm.)-thick plywood was sawed into a hexagonal shape to serve as a base. Small pieces of old oak floor planks were set on the plywood and sawed to shape. When all of the pieces were in place, they were glued and nailed one at a time onto the plywood.

Additional pieces were sawed to make a border round the edge of the table. After sanding the entire table and coating it with a liquid plastic, set it on a base such as this, or any other suitable discard you might have in your attic.

Illus. 240. These trivets may look intricate, but it's actually easy to make them with scrap pieces of decorative moulding.

Pieces of decorative wood moulding left over from a house construction job can be sawed into $\frac{3}{4}$-inch (1.87-cm.)-long pieces and glued together to create a trivet for hot plates. Simply start in the middle and add more and more pieces going in a circular fashion. Glue each piece to the previous piece as you go, and when the circle has reached the desired proportions, tie an elastic band tightly round the outside of the circle. Leave until the glue is dry.

Illus. 241. This unique footstool has old sawed-off golf clubs for legs. Saw a thick piece of wood into whatever shape you like, drill holes in the bottom at an angle and use epoxy cement to hold the legs in the holes.

Illus. 242. This lovely Oriental type of lamp is made of ordinary ice-cream sticks.

Illus. 243.

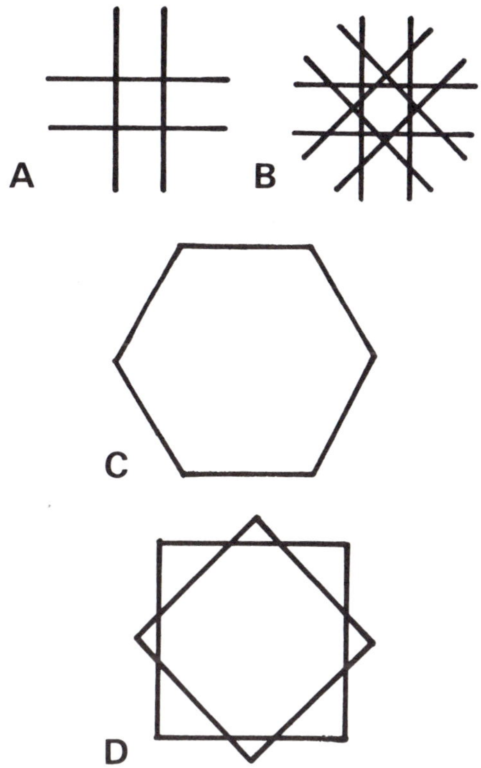

The stunning lamp in Illus. 242 is made of ice-cream sticks! Starting at the bottom, 4 sticks are glued together as shown in drawing A, Illus. 243. Then another 4 sticks were glued in the same way and placed on top of the first 4, at the angle shown in drawing B. Complete this part of the base by making 9 each of the squares shown in drawings A and B. In other words, you will have 18 squares all told.

Next, glue together 8 sticks as shown in drawing C. Make 9 of these, but place them directly one on top of the other rather than at angles to each other.

Then start the next section with 4 sticks glued together with another 4 placed on top of them as shown in drawing D. Make as many squares in this section as you wish depending upon how tall you want your lamp to be.

Finally, make a "platform" consisting of 2 layers of parallel sticks glued to each other as shown in Illus. 243. Drill a hole in this platform to accommodate the light socket and wire. Then, go on to make the top part of the lamp in the same manner as before. You can vary the stacking procedure in any way you like, but be sure the sections are even and in proportion to each other.

Illus. 244. An old wooden bed slat makes an ideal base for a coat-rack. This one was sawed down to 3 feet (1 m.) long. Different-sized wooden thread spools are nailed to the slat for hanging purposes. This rack is painted lavender and decorated with flowers and hearts cut from scrap fabrics, which are attached with Mod Podge. White glue holds the rick-rack trim in place.

Illus. 245. The roof and base of this birdhouse were sawed from old barn wood, and the front, back and sides were sawn from $\frac{3}{4}$-inch (1.87-cm). white pine. To fill the gap between the two parts of the roof, a stick is glued in place. Small pieces of tree bark glued on the front and sides of the house lend a rustic air that is sure to attract birds to the house. The entrance hole was drilled.

Illus. 246. A large piece of driftwood serves as the base for this lamp. A hole was drilled down the middle to hold the metal-tube housing the electrical cord. Tree fungi was glued to the wood for decoration. Then an old lampshade frame was covered with burlap (hessian) and pieces of denim cut from old blue jeans. Pieces of leather and rope complete the decoration.

Glass

Illus. 247. Some of the ways you can decorate glass with glass or with other materials are shown in these attractive glass containers—from left to right a snifter glass, a vase, a vitamin jar and a glass cup.

Although scrap glass is not as abundant as paper, plastic, or some other materials, you will still find that you have access to plenty of raw material in the form of bottles, old mirrors, broken window-panes, marbles, broken vases, pitchers, drinking glasses, light bulbs, jars (what *do* you do with all those baby-food jars!), beach glass, and so on.

In this chapter, you will not only decorate the glass itself, but use pieces of glass to decorate other objects. You will also actually cut glass to make your own candle holders, and more.

The unusual vases and containers in Illus. 247 are an example of how easily you can create with glass. On the left, a glass snifter is covered with colored glass beads, sequins, imitation gems, and Christmas glitter. White glue holds everything on.

The tall glass vase was smeared with glue, and colored aquarium gravel was applied in an interesting pattern. Next to the vase is a vitamin jar covered with shell macaroni, which was glued on and then sprayed with gold paint.

The glass cup on the right is decorated with colored glass chips and pieces which were glued on with tile cement. Solid silver which comes in a tube was applied to create a mosaic effect. This silver takes approximately 3 to 4 hours to dry. This cup would make a perfect candle holder.

Illus. 248. In this mosaic plaster of Paris forms the "grout," the white areas between the pieces of colored windshield glass.

Mosaics of various kinds can be made easily with broken glass. You can either use glass that is already colored, for example from a stained-glass window or marbles, or plain window glass that you can color with glass stain before breaking it up. When making glass tesserae, place your piece of glass in a burlap bag, put on gloves, and with a hammer, smash the glass through the burlap. In this way, there will be no danger of flying glass or glass fragments being left on your working surface. Remove the glass from the bag carefully with your gloved hands, and sort out the pieces according to size, shape, or color. Keep your odds and ends of shattered glass in small boxes for future use in your glass scrap crafting. You can glue your glass tesserae onto plain glass, a mirror, or directly onto a window-pane, or other surface.

For example, broken windshield glass provides the tesserae for the circular mosaic in Illus. 248. The glass, before it was shattered as described above, was painted with various colors of glass stain. Then it was smashed and the colors sorted and glued to a circle sawed from plywood. A pencil design was drawn on the plywood to help place the tesserae in a pleasing design.

Then, plaster of Paris, mixed with water to the consistency of mashed potatoes, was applied over the tesserae and pushed into the cracks between them. The excess plaster was washed off immediately.

Illus. 249. Here, broken windshield glass was glued to light green wall panelling. Some areas of the glass were painted with varying shades of green acrylic paint before the glass was smashed. Blue yarn separates the areas of the design.

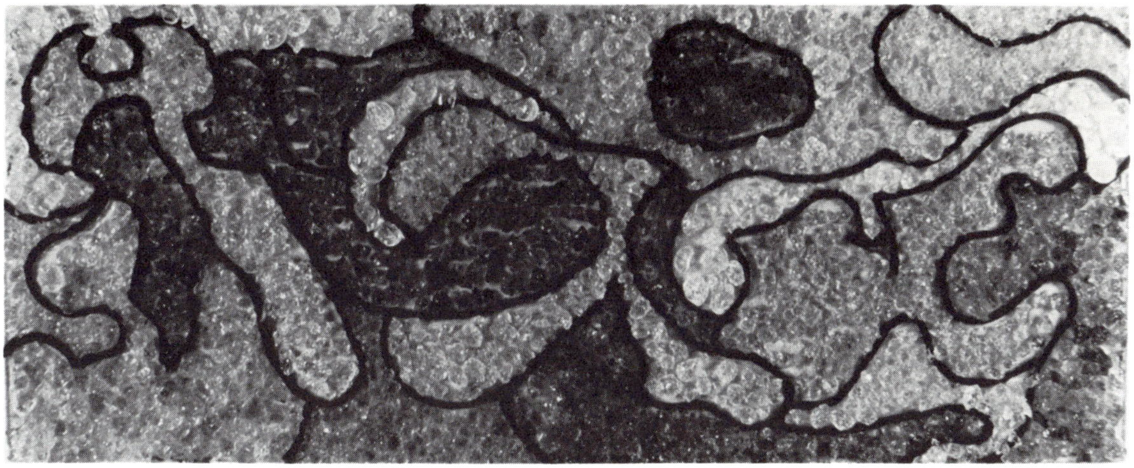

Illus. 250. Here's an idea for putting all those chipped marbles to use—a crazed-marble mosaic!

The tesserae in Illus. 250 was created from chipped marbles that were subjected to a process called "crazing." Place your glass marbles in a sieve and lower into a pot or pan of boiling water. Make sure they get piping hot and then remove them one by one with a slotted spoon and drop into ice-cold water. They will shatter immediately. If they don't, the water is not cold enough. Either try again, or place the unshattered pieces in a bag and smash with a hammer. In this mosaic, yarn and cord were first glued in a free-form design onto a clear glass background before the marble bits were glued on.

Illus. 251. Colored glass bottles were crushed in burlap bags with a hammer to provide the tesserae for this mosaic on a plain glass background.

Illus. 252. The three bottles on the left were decorated with painted glass, while the one on the far right boasts a covering of yarn, seeds, and aquarium gravel. All are attached with white glue.

The three bottles on the left in Illus. 252 were decorated with painted glass. First, place your clear glass on newspaper and then paint different colors of glass stain on it. Apply plenty of color so that the coating is not thin. Let dry and then smash in a burlap bag. Paint your bottles with white glue. Then make a 1-inch (2.5-cm.)-thick line of glue from top to bottom. (Use plenty of glue.) Press the pieces of fractured glass tightly against the bottle to ensure against air spaces. Next, add another line of glue and glass, and so on, until the bottle is covered. Wipe off excess glue and let dry. Apply plaster of Paris in the cracks. You can color the plaster-of-Paris grout if you wish when it is dry.

Illus. 253. Torn and cut pieces of tissue paper were glued to the middle of this wine bottle with white glue diluted with water. Then the top and bottom of the bottle were smeared with glue and black yarn was wound tightly round the exposed glass, starting from the bottom. The criss-cross lines of yarn were glued on last.

Illus. 254 (above left). This tall bottle was smeared with glue and heavy, colored cord was wound tightly round it. When doing this, work in sections, starting from the bottom. When the glue is dry, glue pieces of eggshells to form "ruffly" borders. To make the bouquet at the top, poke holes in eggshell halves with a large needle and insert pipe cleaners to act as stems.

Illus. 255 (above right). Here, coiled pieces of yarn were glued onto a wine bottle to simulate flowers. Black yarn was glued on in a line design which allows the green of the bottle to show through as part of the over-all design.

Illus. 256 (right). This wine bottle was covered completely with yarn. When making a design of this kind, plan it on paper first.

Illus. 257. A large wine bottle is here decorated with fabrics cut into ovals and glued on. Bias binding tape was glued on for added interest.

Pop-bottle dolls are not only fun to make, but the results are often spectacular as you can see in Illus. 261 to 264. You can use either a tall pop bottle or ketchup bottle for your dolls. First, wrap a piece of absorbent cotton (cotton wool) in a square cut from a nylon stocking. Gather the nylon round a pipe cleaner and insert it into the mouth of the bottle. You now have a flesh-colored head. Then wind a long pipe cleaner round the neck of the bottle once and extend the ends out to the sides to form a foundation for the arms as shown in Illus. 258.

Next, wind toilet paper round the arms, chest, and back as loosely as possible to create a full-bodied effect (Illus. 259). A long piece of paper, folded once, provides the right height for the chest and back. Hold the paper in place with elastic bands.

Crepe paper, because it is so easily stretched and handled, provides the ideal material for clothing for your doll. In Illus. 260 a crepe-paper skirt is being shirred round the waist of the doll and secured with an elastic band. Cylinders of crepe paper, also held with elastic bands, have been added in Illus. 261. Rectangles of crepe paper glued to the front and back complete the garment. Pieces of ribbon, sequins, and yarn hair add the finishing touches. This old-fashioned doll also had toilet paper stuffed into the back of the skirt to create a bustle.

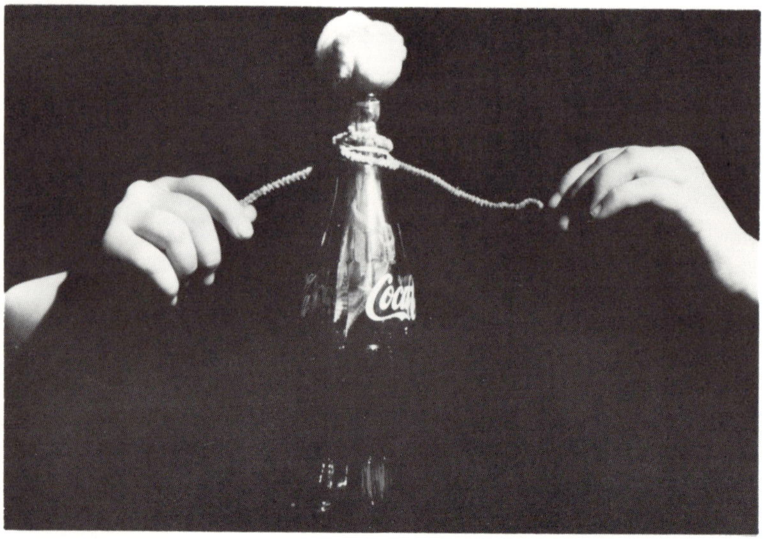

Illus. 258. A tall pop bottle or ketchup bottle is the perfect foundation for dolls and other figures. Here the doll's head is in place and her arms are under construction.

Illus. 259.

Illus. 260.

Illus. 261.

A rum bottle provides the foundation for "Caesar" in Illus. 262. His toga is made from a piece of an old sheet and other scrap cloth. His face is made of clay, which was modelled over the nylon and absorbent-cotton head. Next to him, in Illus. 263, stands "Napoleon" in a uniform made from colored construction paper. The Japanese geisha doll is made over a pop bottle. A piece of flowered cloth was used for her costume. Her hair is made from unravelled yarn and her facial features are painted on with India ink.

Illus. 262.

Illus. 263.

Illus. 264.

Illus. 265 (left). These candle holders were made from two beer bottles, cut with the glass cutter shown in Illus. 267.

Illus. 266. You can make wind chimes if you cut off the tops of pop bottles with the glass cutter.

Once you learn how to cut glass, you will be able to make all kinds of interesting objects with your scrap pieces and throwaways. Although there are commercial bottle cutters available, if you have a glass cutter such as the one shown in Illus. 267, you can use it for both bottles and flat pieces.

Wrap a piece of masking tape round a bottle where you wish the cut to be. Use the edge of the tape as a guide and score the bottle with the cutter. Then, hold the bottle by the two ends and rotate it slowly over the flame of a candle, at least 3 or 4 times, directly along the scored line. *Do not let the glass touch the flame.* Then quickly rotate it another 3 or 4 times and run cold water along the score. Tug gently on both ends and the bottle will pull apart. Take emery cloth and smooth down the edges.

To make the candle holders in Illus. 265, melt down old candles and pour the wax into the bottom halves of two bottles. Set thick candles

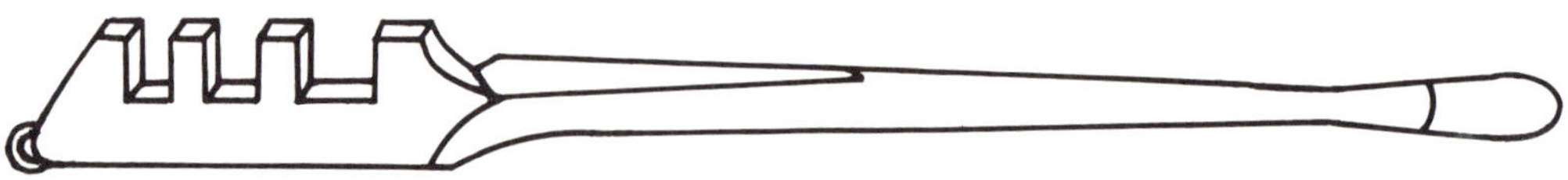

Illus. 267. Glass cutters have a small wheel at the tip that does the actual scoring of the glass. The notches are for breaking off small pieces or rough edges.

Illus. 268. The two candles here were cut with the glass cutter from a broken, red, stained-glass window, and glued with epoxy cement to a sheet of plain glass. (Follow the manufacturer's directions for cutting flat glass.) Gold foil paper was glued onto the back and the candles surrounded with small pieces of stained glass and plastic foam packing material.

into the bottom. Use the top halves to make the wind chimes in Illus. 266.

The top sections of five pop bottles were cut as described above to create the wind chimes. Wire coat hangers were cut into 3-inch (7.5-cm.) lengths and bent into a U shape to fit inside the glass chimes. The wires are tied to a hanging cord that ends at the bottom with wooden beads and a small square of brass. Pieces of felt decorate the tops of each of the chimes. Hang the chimes in a breezy window or hall.

Illus. 269 (left) and 270 (right). A gallon-size glass apple-cider jug provides the base for this colorful lamp. The jug is filled with peas and beans of different colors. Yarn and artificial fruit decorate the lamp. A light socket was inserted into the mouth of the jug, and the burlap covered shade was decorated with corn kernels, dried flowers, and yarn.

Illus. 271 (left). A half-gallon-size whiskey bottle is the base for this interesting lamp. The bottle was left undecorated and a tan burlap (hessian) lampshade was added with decorative trim that matched the colors in the bottle.

The towering candlestick in Illus. 272 is composed of a variety of glass containers. Since it was too tall to photograph, the upper four pieces are shown on the right. They were glued on later. The glasses were arranged so that the heavier ones are on the bottom and the lighter on top and so that the bottoms and tops would fit each other easily.

At the bottom is a glass ash tray filled with plaster of Paris. The glass containers were glued together two at a time and allowed to dry. Then, they were combined to create sections of four. When they were dry, the fours were glued together. Some of the pieces were painted, while others had designs made of scrap materials glued on them. Inside some of the glass containers a variety of dried flowers, sand and seashells was placed. Finally, a candle was set in the top wine glass.

Illus. 272. "Seasons of the Year" is the title of this candlestick composed of mustard jars, baby-food jars, wine glasses, and so on, some of which are filled with natural materials suggestive of the seasons.

Metal

Illus. 273. This stunning portrait is made of two different thicknesses of wire twisted by hand. The two ends are nailed into a piece of scrap wood which acts as a base. The eye was cut out from a piece of paper and glued to the wire.

Scrap metal is widely abundant in a great variety of forms—from tiny pins to old cars. Some suggestions for raw material include nails, wire, aluminum pie plates, tacks, old belt buckles, flashlights, nail files, broken scissors, knives, forks, spoons, pots, pans, tin cans, pieces of pipe, nuts, bolts, washers, used parts of such appliances as washing machines, dryers, toasters, broken typewriter parts, used car parts, bicycle parts, broken toys, ballpoint pen cartridges, paper clips, aluminum foil, copper foil. Most of these items are right in your own home, but don't overlook the local junk-yard, car repair shop, or even your friends' trash piles!

Some of the ideas in this chapter require nothing more than glue, while others will test out your skill at soldering and brazing. On this page you will start out with simple wire creations that a child can do.

Illus. 274. Thin wire twisted together forms these happy little dancers who are attached to a piece of driftwood with staples. Before the figures were attached to the wood, melted wax from colored candles was dripped over the wire.

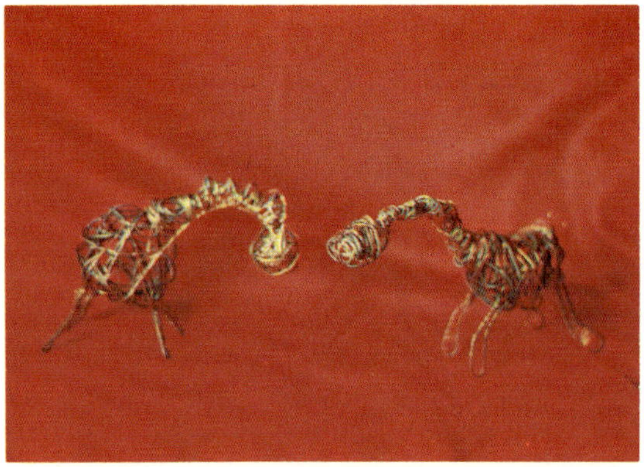

Illus. 275. These two giraffes were made by wrapping wire round and round simple wire skeletons.

Illus. 276. This beer-can coffee table is made from 78 cans glued together with epoxy. A mirror is glued to the tops of the four walls and a fringe glued to the mirror for decoration.

You need no more than glue to make the beer-can coffee table in Illus. 276 and the pop-can house in Illus. 277. The base of the house is made of 9 cans glued together and wrapped with tape until the glue dried. The construction-paper roof is supported by 6 cans placed on an angle,

Illus. 277. Tin cans form the base and the supports for the roof of this charming little house. Construction paper is glued on for the windows, door and chimney.

Illus. 278. This tin-can robot is made up of a variety of cans.

Illus. 279. For this lamp, you will have to cut some cans with a tin snips.

3 on each side and in turn supported by 2 cans placed flat.

Illus. 278 shows a robot who is made up of both whole and cans cut with tin snips. Wire, inserted into the body supports the arms. Can lids make up the feet and eyes, and part of a can with slits cut into the rim forms a jaunty hat. The robot's nose is a plastic-foam ball glued in place.

The shade of the lamp in Illus. 279 is made up of beer cans which are cut apart along the seams and then flattened with a rubber mallet. They were then glued onto an old lamp shade, and fringe was added for decoration. The base consists of 9 cans joined with epoxy cement. A hole was drilled through a square piece of wood and glued to the cans, allowing for the light cord and tubing to pass through.

Illus. 280. To create this bas-relief, sections of aluminum fruit and meat plates were cut and glued with epoxy to roof-nail heads driven into a piece of plywood which was painted black.

Illus. 281. Driftwood and nails joined together with epoxy cement create this table decoration.

A piece of driftwood provides the base for the table decoration in Illus. 281. Some of the "flowers" consist of nails joined together with epoxy cement and others of the tops of pop cans cut with a tin snips to simulate petals. All of the "stems" are nails driven into the wood.

Furnace filters were used for the framework of the hanging lamp in Illus. 282. First, a piece of tin approximately 4 inches (10 cm.) wide was shaped into a cylinder approximately 5 inches (12.5 cm.) in diameter. The ends were joined with a small screw. Then, a section was cut from a furnace filter (you can use any similar filter) and glued round the tin cylinder. A second filter, the same size as the first one, was glued on the outside, providing a double thickness. The border at the top consists of black velveteen and gold braid. Tiny beads hang from thin wire tied to the tops of the filter openings. Four holes were punched into the tin at regular intervals to accommodate the hanging chains which were removed from an old brass planter.

Illus. 282. This hanging lamp was made from furnace filters.

Illus. 283 (left) and 284 (above). This tall metal ashtray consists of a car cam shaft and two iron castings, one on each end of the shaft. The bottom casting is glued to the shaft with epoxy cement, while the upper casting simply fits into the top of the shaft so that it can be removed for cleaning. By sheer accident, the pieces fitted together perfectly!

Illus. 285. Metal computer-tape roll covers, painted orange, created these hanging candle holders. Twine is inserted into four holes drilled in the rims for hanging. Beads of various kinds decorate the twine and the holders. Short votive candles keep the flame away from the twine.

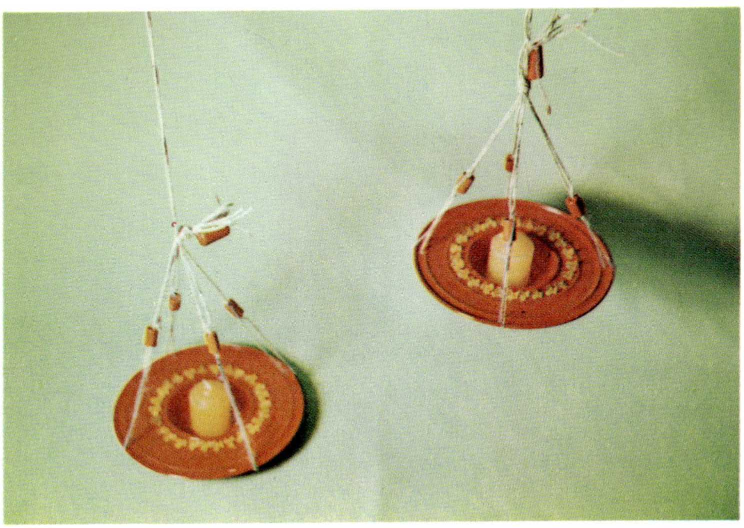

Illus. 286. Parts of a metal toy construction set and model plane parts are the ingredients for making this unique airship.

The fanciful plane in Illus. 286 has a base made up of parts of an old toy Erector set, including the wheels and platform. An old-fashioned dating stamp is attached to the platform. The wings and body are made from a discarded plastic watch display holder, and the remaining parts all come from old model plane sets. Nuts, bolts and glue are the joining materials.

An antique kitchen sink pump is the body of the lamp in Illus. 288. The pump is screwed with machine screws to a piece of barn wood. A vertical rod, housing the electrical cord, passes through the middle of the pump, supporting the socket and harp for the shade. Burlap was sewn on an old paper shade and trimmed with yarn and old photographs. The pump was cleaned with steel wool and painted with antique orange paint. Then a glaze was applied and rubbed lightly with cheesecloth. Finally two coats of satin varnish were applied.

Illus. 287.

Illus. 288.

Illus. 289. These unusual candle holders are made from two cans each.

Each of the group of four candle holders in Illus. 289 is composed of two cans glued together end to end. Parallel slits, about $\frac{1}{4}$ inch (.6 cm.) apart, were cut in both ends with tin snips and curled up over a knitting needle to form the bottom supports and the decorative top parts. Candles with their wicks were melted into the holders.

A set of miniature chairs like the one in Illus. 290 would be ideal for a doll house. Only half a can with its bottom intact is needed for one chair. Closely spaced slits are cut perpendicular to the cut rim. Some of the strips were then cut off, while other strips in the back were bent straight up to provide a back for the chair. The bottom of the can then becomes the seat. The remaining strips in the front were curled up in a decorative fashion. The back is made by tying a small strip of metal round the bent-up strips so they are held tightly together as shown. Then the ends of the strips are curled and glued to each other. A tiny corduroy cushion completes the chair.

Illus. 290. A miniature chair made from half a can makes an elegant addition to a doll house. Many other items of tiny furniture can be made in the same way.

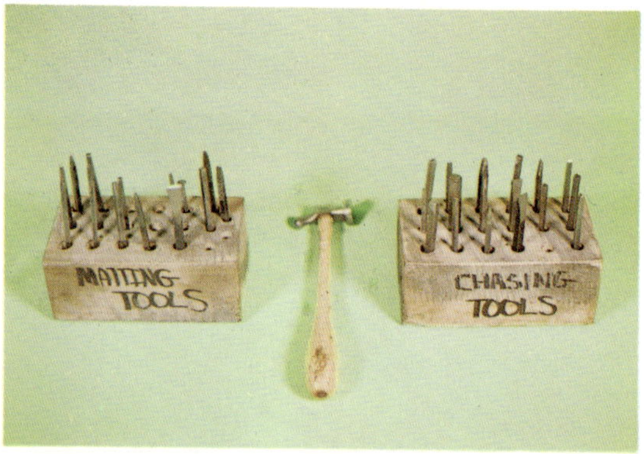

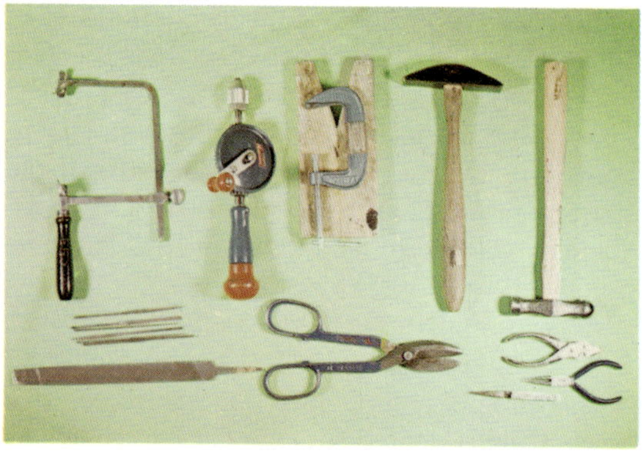

Some tools you might need in your metal crafting. Illus. 291 (left): from top left to right: jeweler's saw, hand drill, C-clamp and bench pin, forging hammer, ballpeen hammer, files, tin snips, pliers and tweezer. 292 (right): matting tools, ballpeen hammer, and chasing tools. Matting tools are punches with different head shapes which are placed over the metal and pounded for attaining a relief effect. Chasing tools define and texture the metal after the relief is achieved or can be used on flat surfaces.

Now that you have learned some of the simple techniques of cutting and joining your metal creations, you may wish to try some more advanced methods such as chasing, forging, matting, soldering—all of which require certain pieces of equipment. A number of tools are shown in Illus. 291 and 292 which can be used for the projects on the following pages.

The rectangular pieces of metal forming the wind chimes in Illus. 293 were sawed from a large sheet of scrap copper. Then the copper was placed on a heavy mat and the design produced by striking a matting tool with a hammer. Holes were drilled all the way through a piece of driftwood and at one end of the copper strips to accommodate the copper wire which acts as the support for the chimes.

The "sun god" in Illus. 294 is composed of copper wire, a washer, a bolt, and a piece of copper with a stone glued on it. First, a piece of copper wire was shaped into a circle. Next, pieces of copper wire were hammered until they were flat. Use the flat face of the hammer for this, holding the metal firmly on your work surface. After striking one side for a while, turn the piece over and strike the other side in

Illus. 293. This gleaming wind chimes will give you a chance to use your matting tools and hammer.

Illus. 294. Flattened copper wire produced the rays of this "sun god."

exactly the same place with the same force. (This is called cold forging.) One end of each of the sun rays was bent round the circular piece to fasten it in place. The circle and the nose and eyes are held in place with bent pieces of wire hammered into the wood. The ends of the rays are held flat with glue.

The interesting neckpiece in Illus. 295 is composed of copper wire and a piece of flat copper. Pieces of wire were flattened on both ends with a hammer. Small pieces of wire were then cut and shaped into small circles with a pliers (you can use tweezers, too) and inserted into holes drilled in the flattened ends of the other pieces. Chasing tools were used to decorate the large copper piece, which then had a number of small holes drilled into its top edge. Tiny copper circles strung with beads were inserted into the holes.

Illus. 295.

Black burlap, glued over a piece of plywood which was cut to shape, is the background for the aluminum ship in Illus. 296. The aluminum scraps, cut from tin cans, were shaped over a mould (a car hubcap was used here). Nail holes produced the portholes. The bowsprit and masts are made of wire cut from a coat hanger.

Illus. 296. You can easily shape pieces of aluminum to create this sailing ship.

Illus. 297. Heavy copper wire and wooden beads are the only materials you need for this neckpiece.

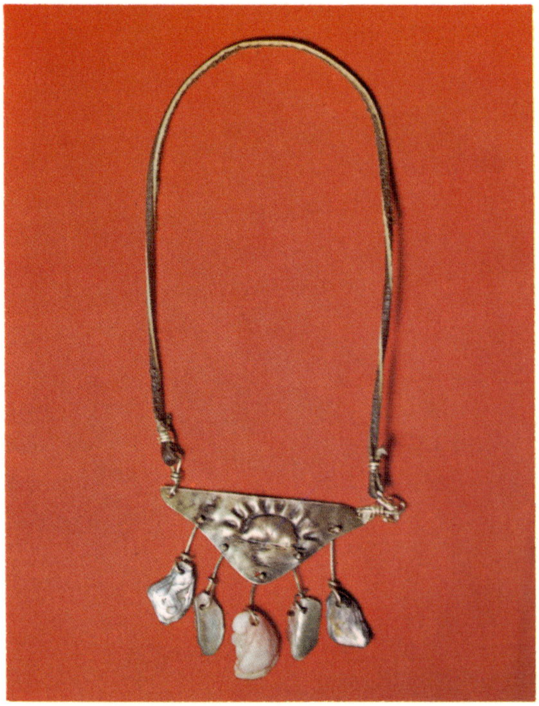

Illus. 298. "Fort Lauderdale Sunrise" is the title of this found-object and metal necklace.

A long strand of copper wire serves as the neckpiece in Illus. 297. The ends are bent into a U shape with a thin-nosed pliers to form the clasp. Short pieces of wire flattened with a hammer are attached in the same manner to the strand. The wooden beads were strung before the wire was bent.

The sunrise design on the necklace in Illus. 298 was achieved through chasing (see Illus. 299). Small holes were drilled along the bottom of the triangular piece to support brass wires, the ends of which are inserted through drilled holes in seashells. The ends of the wire were heated with an acetylene torch so that the brass melted into balls, preventing the wire from slipping out of the holes.

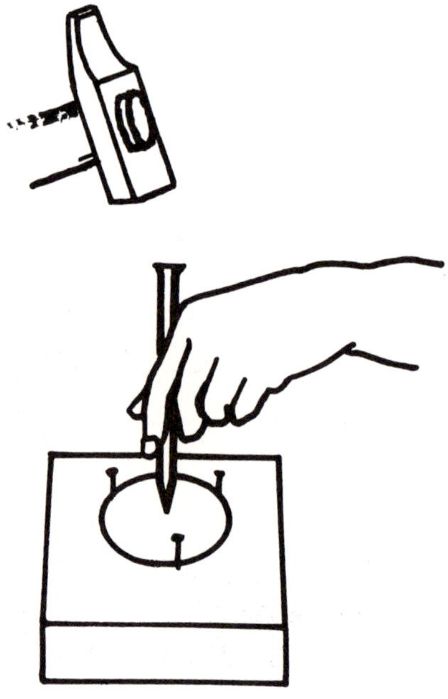

Illus. 299. Chasing.

Illus. 300. Materials needed for soldering include a propane torch, silver solder, emery paper and flux.

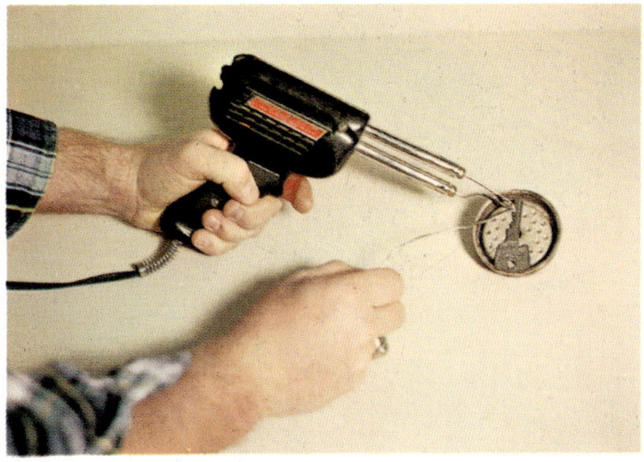

Illus. 301. A soldering iron and acid-core wire solder can be used in place of, or in addition to, a propane torch.

The suggested projects on the following pages were all created by a process called soldering, which will allow you to join metals together in such a way that you can make certain objects that you would not be able to do otherwise. You will need either a soldering iron or an inexpensive propane torch. Disposable cans of pressurized propane gas are available for use with a reusable torch attachment.

Metals to be soldered must be perfectly clean. Buff away with steel wool all paint, rust, oil, varnish, lacquer, grease and fingerprints. If necessary, burn away stubborn materials with the propane torch and polish the residue with steel wool. A wire brush and a strong solution of detergent and hot water will remove greasy and oily films. Emery paper or moist pumice brushed on with a bristle brush are also effective.

When using a soldering iron, brush the tip with a wire brush until it is shiny. Apply commercial liquid or paste flux to the metal joint to be soldered and to the tip of the iron. Melt a droplet of solder onto the tip to "tin" it. Then heat the metals with the iron. (Be sure to work over an asbestos pad.) Apply a stick of solder to the joint until a bead is melted away. Move the iron over the metal until the solder flows smoothly. The surfaces must fit together well if the solder is to work properly.

Solder and flux are required for soldering. Acid-core soldering wire is most useful for joining scrap metal objects. Acid flux must be applied with a brush to prevent oxides from

Illus. 302. This car was created through a soldering process called "brazing," in which unlike metals are joined. When like metals are joined by soldering, it is called "fusing."

Illus. 303. Paper patterns were first made for this mobile. The shapes were sawed out of 16-gauge galvanized sheet metal, and then spray-painted. The supports are provided by a cold rolled steel rod soldered to the large metal circle and cold copper welding rods which support the small metal discs. Holes drilled in the large circle lighten its weight.

forming. Apply the flux to the top side of the metal. Be sure to use plenty of flux on copper as it oxidizes quickly. Scrub fluxed metals in soapy water immediately to prevent discoloration. A second bath in a weak solution of vinegar and water will act against any remaining traces of acid. Flux is very corrosive, so handle with care.

Soldering with the propane torch is much the same as with the soldering iron except that the metal is first heated with a medium-hot flame. Next, it is fluxed and reheated. The solder is then dabbed against the hot metal to leave a deposit of small beads. Do not apply the solder directly to the flame. Play the flame over the joint just long enough to completely liquefy the solder. The propane torch is inexpensive, and the intensity of the flame can be controlled by turning a dial. Although soldering wire is often

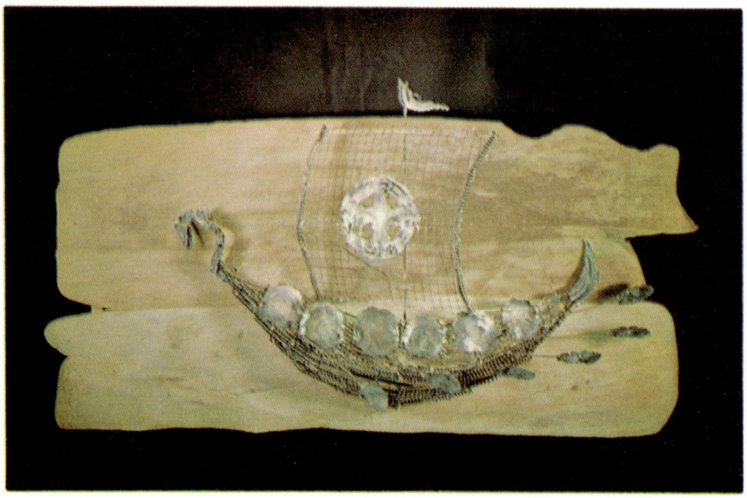

Illus. 304. A model was first made of this Viking ship with pipe cleaners. Tin can parts soldered together were used for all parts of the metal ship, except the sail which is made of hardware cloth. The soldering iron was applied to some of the parts to give them an antiqued appearance. The finished ship was screwed to pieces of driftwood.

Illus. 305. This sporty bicycle was created from copper welding rods, steel washers and sheet metal. All parts were brazed together with melted brass.

used, you can also use the type of solder shown in Illus. 300. With a tin snips, you can cut this small sheet of silver solder into very small pieces which are called paillons. These should be about $\frac{1}{32}$ inch (.79 mm.) long.

Sweat soldering involves soldering together two or more layers of metal so that they are laminated. Solder is fused in between the layers and not only round the edges. The metal parts are placed on an asbestos pad with the surfaces to be soldered face up. Flux is brushed over the top layer and the metal is heated with a medium high torch flame until hot. The tip of a stick of solder is dabbed on the metal so that small dots of solder appear. More flux is applied and the torch flame is run over the solder until it flows into a thin, even film.

The bottom layer of metal is heated, fluxed, and the top layer is placed over it with its tinned side down. The torch is applied over the metal sandwich until the solder is melted. Hold the laminated sandwich in place with the point of a thin-nosed pliers so that the laminations do not become misaligned during the soldering. Car keys, clock parts, washers, and other articles can be soldered to a flat metal surface in this manner.

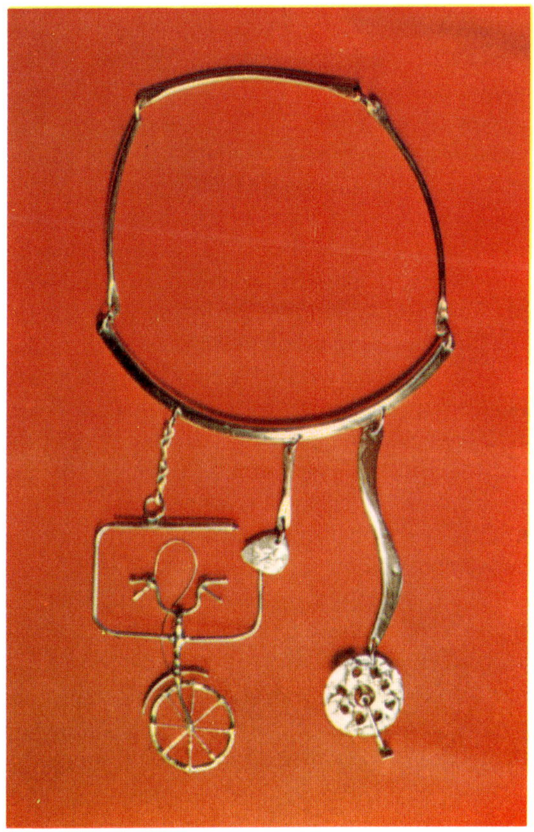

Illus. 306. Cooper tubing, brass, and copper wire were here brazed to form the bicycle-like arrangement on this unique necklace.

Illus. 307. Old telephone bells form the playing pieces for this checkers (draughts) set. Discarded carpeting and wood moulding are all you need for the board itself.

An old wooden table top serves as the base for the checker (draughts) board in Illus. 307. A few inches inside the edge of the table, four 16¼-inch (40-cm.)-long pieces of wood moulding were glued to the wood to form the outline of the checker board. Outdoor carpeting in two different colors was cut into 2-inch (5-cm.) squares, 32 of one color and 32 of the other. Starting in one corner, alternating colors were glued into place within the moulding.

The 24 playing pieces were created from old telephone bells. Two bells were rivetted together with a copper rivet to form each playing piece. A 20-gauge brass plate was silver-soldered on the top of each piece and a copper plate was soldered on the bottom. When a player acquires a king, he simply turns the piece over so that the copper plate is on top.

All of the pieces were buffed, and then 12 pieces were oxidized to a golden brown, to provide the second color. They were dipped into a hot solution of one part copper sulfate and two parts water. The longer the dipping, the browner the pieces will get. For an even color, brush the pieces with pumice and water after the dipping. A second dip in the solution will then provide a more uniform shade.

Illus. 308.

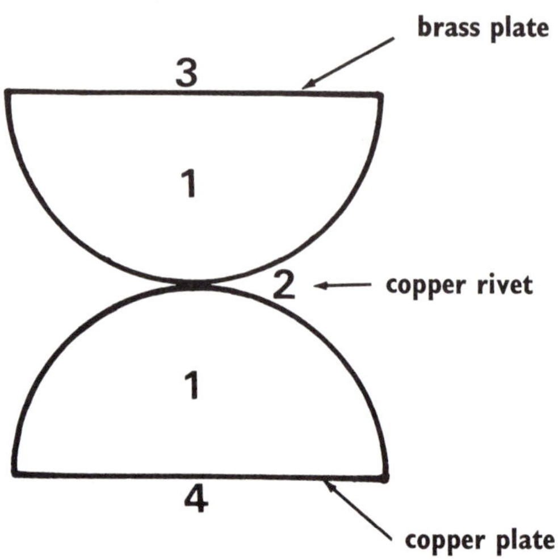

The table top and moulding were walnut-stained.

About the Author

Chester Jay Alkema was born and raised in the rural village of Martin, Michigan, and attended the Martin Agricultural High School. He graduated from Brandon High School, Brandon, Wisconsin, in 1950; received his A.B. degree in Education from Calvin College, Grand Rapids; earned his M.A. in Art in 1959 from Michigan State University, and his Masters of Fine Arts degree in 1961.

Prof. Alkema has taught art to children of the Seymour Christian School, Grand Rapids, Michigan, and the Wyoming Parkview School, Wyoming, Michigan, for many years. Since 1959 the author has taught adult art education courses for Michigan State University in East Lansing, Grand Rapids, Benton Harbor, Battle Creek and Niles. In the Spring of 1965, he joined the art faculty of Grand Valley State College, Allendale, Michigan.

Prof. Alkema has contributed numerous articles and photographs in the national educational journals, *Arts and Activities, School Arts, The Instructor, Grade Teacher, Design, Exceptional Children, Children's House*, and the foreign publication *Saber*. He has illustrated stories for *Crusader*, a magazine for children. This book is the thirteenth book written by Prof. Alkema, twelve of which have been published by Sterling since 1967.

Index